Miners and Medicine

Miners and Medicine
West Virginia Memories

By Claude A. Frazier, M.D.

With F. K. Brown

Foreword by Stuart McGehee

University of Oklahoma Press : Norman and London

Other Books by Claude A. Frazier, M.D.

Coping with Food Allergy (New York, 1977, 1985)
Insects and Allergy, and What to Do about Them (with F. K. Brown)
(Norman, 1979)
*Coping and Living with Allergies: A Complete Guide to Help Allergy
Patients of All Ages* (Englewood Cliffs, N.J., 1980)
*Insect Allergy: Allergic and Toxic Reactions to Insects and Other Ar-
thropods* (St. Louis, 1987)

To Beatrice Bryson, a true friend

Frazier, Claude Albee, 1920–
    Miners and medicine : West Virginia memories / by Claude A.
Frazier, with F. K. Brown.
        p.   cm.
    Includes bibliographical references and index.
    ISBN 0-8061-2454-7 (alk. paper)
    1. Coal miners—Medical care—West Virginia—History. I. Brown,
F. K. (Frieda Kenyon), 1921–   . II. Title.
    [DNLM: 1. Coal Mining—history—West Virginia. 2. Occupational
Diseases—history—West Virginia. 3. Pneumoconiosis—history—West
Virginia.   WA 11 AW4 F8m]
RC965.M5F73   1992
616'.008'8622—dc20
DNLM/DLC
for Library of Congress   92-054151
                                                                       CIP

The paper in this book meets the guidelines for permanence and
durability of the Committee on Production Guidelines for Book Lon-
gevity of the Council on Library Resources, Inc. ⊚

Contents

Illustrations

Miners and Medicine

Foreword

BY STUART McGEHEE

"Is that urine?" asked the bespectacled coal-company doctor as a shy patient held out a specimen cup. "No sir," replied the sturdy miner. "It's my wife's."

Claude A. Frazier and F.K. Brown's *Miners and Medicine* is the first book-length study of a long-neglected and fascinating aspect of American history. Before the arrival of the railroads and the opening of coal mines in the late nineteenth century, Appalachian health care was a morass of quackery, elixirs, and bucolic superstition. Mountaineer "physicians"—medicine men, really—were not often graduates of accredited medical schools. Community health care was nonexistent, woeful, or downright dangerous in communities where smallpox victims' open-casket funerals, for example, could result in widespread family epidemics.

All that changed swiftly with the coming of the steel rail and the coal mine. The isolated and scattered mining communities required more systematic, modern health care. Labor was frightfully scarce, and trained doctors were an incentive to attract and keep good miners. The sudden and unprecedented growth of the region created the need for the company doctor, one of the most characteristic and storied institutions of coal culture.

The company doctor, charged with the health care of an

ix

entire community, could not afford to be a narrow specialist. Rather, he was the epitome of the American country doctor. Beloved coalfield doctor Bernard Clements, of Matoaka, West Virginia, delivered a whopping 3,500 babies before his untimely death in 1971 at the ripe old age of ninety-one.

Claude Frazier, a McDowell County physician's son who followed in his father's footsteps, tells this story with respect and affection, with the help of F. K. Brown. At the heart of the narrative are dozens of anecdotes sent in by coalfield residents at Doctor Frazier's request. In their own words, coal people poignantly describe the importance of the coal-camp doctor in their lives. More than anything else, these memoirs reflect the human side of health care in a turbulent and often misunderstood industry. Frazier and Brown have performed a remarkable public service by uncovering the unique story of miners and medicine.

Physicians are responsible for that most precious of commodities, human life. Coalfield doctors understood that obligation as a commitment and practiced their special art with a diligence and a dedication that befitted the medical profession's high ethical standards. Hippocrates would be proud.

Acknowledgments

We wish to extend our gratitude to all those who answered our request for information about the coal camps of yesteryear. We received well over one hundred letters, clippings, and pictures, every one of them helpful in the writing of this book. Although we could quote from only the sampling listed in the bibliography, they all provided valuable descriptions of coal-camp life in West Virginia and other areas of Appalachia, and of what it was like to be a coal miner or a member of a miner's family, or the company doctor responsible for the health and welfare of a coal camp. We thank all our correspondents.

We would also like to express our gratitude to Dr. John Duffy and Dr. Richard P. Mulcahy for their valuable help in the final preparation of our manuscript.

CLAUDE A. FRAZIER
F.K. BROWN

Miners and Medicine

Prologue

Not long ago, I stood in the middle of a deserted coal camp and stared at the white house which once had been my home. The son of a coal-camp doctor, I had spent a good deal of my boyhood in this mining community. There once were small houses along the narrow, dusty road; not six feet apart as I remember them. They were stained with coal dust, but graced by green lawns and flowers. My father's house was larger, for the company doctor rated up with mine officials when it came to company-owned housing.

It is strange to go back to what was once a teeming, crowded community only to find that most of it has vanished. Only a few of the miners' houses are still standing, and those are so spruced up and added onto that they bear little resemblance to the original. Most of the camp has been torn down or stands crumbling and empty, the vegetation slowly encroaching. Down the road beside the rusting and weed-choked railroad tracks, the tipple, once the scene of so much activity, is falling to ruin. The chutes containing the conveyor belts, which brought the coal from the mine portal high on the mountain, stand silent behind their warning signs. No rattle and clank of coal cars, no grumbling of the donkey engine moving the conveyor belts, no clatter of coal pouring down from the portal into the waiting railroad cars. The valley is silent now.

An abandoned tipple. *(All illustrations are reproduced courtesy of the West Virginia State Archives, Charleston)*

The contrast was striking between this melancholy, almost spooky place and my memories of the dusty street when it was so filled with people that it often seemed that a parade was in progress. It was as if I had only dreamed of playing in the coal-dusted yard, of climbing among the coal cars, of laying my pennies on the tracks to be flattened by the passing coal trains. Yet I almost could see the miners' wives heading with their baskets and their pokes toward the company store, where off-shift miners and the retired or disabled sat on the wooden porch, whittling and shooting the breeze. At the end of a shift miners came down through the town, their faces and clothing black with coal dust, their empty lunch pails swinging. In those days the mineowners had yet to furnish a bathhouse where miners could shower and change into clean clothes. A big galvanized tub at home was used to transform the coal-blackened into the clean, and many a miner's wife had to expend considerable effort to restore her husband. Even as a

Abandoned mine

boy, I was aware of the weariness in their steps as the miners clumped along the street. Perhaps I sensed their relief at having survived one more day under the mountain.

There was always movement in the valley then. Coal cars were constantly shunted back and forth by a puffing engine. I clearly remember the whistle blasting for the change of shifts—and sometimes to warn of a potential emergency deep in the mine. Now there is only bird song, the whisper of the wind in the secondgrowth trees on the mountainside, the chuckle of the creek running its rusty red course, stained by acid leaching from the slag pile beyond the camp.

Life in the West Virginia coalfields was hard. It is odd that so many of the miners, their wives, and children have written me that they look back with pleasure on those days of relatively miserable living conditions, of difficult and dangerous work, of simple, spartan play. Perhaps this is chiefly because of the intense sense of community that existed in most of the camps. Miners and their families were isolated because roads, where they existed, were poor and the camps were situated in narrow, hard-to-reach valleys and coves. They had to rely on each other, in joy and sorrow, and especially in sickness and

5

injury. The miners and their families helped each other, for they were in a sense, one big family. They were in it together, sharing the same anxieties, the same tragedies and pleasures, solidly allied against the company supervisors and the wealthy mineowners.

The Appalachian coalfields have bred grinding poverty, appalling tragedies, and violence—the legacies of the capitalist system operating at its worst. There is no pride in King Coal's profits, for they were generated at the expense of the lives and health of thousands of miners. From the time coal was mined in quantity until World War II, there were few bright spots. Instead, we see unrelieved misery in the exploitation of human beings for immense profit for a comparative handful of owners and stockholders, most of whom lived a good life far from the mountains and tight valleys of Appalachia. The comforts that did exist came from a few enlightened mine operators, and from what the miners and their families themselves could create.

Many good books have been written about the coal miner and the coal-mining industry from the standpoint of technology, unionization, and the like. This book combines the two great influences of my life: coal and medicine. Coal-company doctors such as my father brought the science of medicine to the isolated coal and lumber camps as well as surrounding rural areas. Coal-company doctors also initiated the building of area hospitals.

Not all company doctors were good people; some were lazy and incompetent. But I believe it is safe to say that the majority were conscientious people of courage and great endurance who did the best they could for their patients under conditions that would try the souls of today's physicians. These traits made them a perfect match to their patients, the miners and their families who bravely endured. King Coal was not a benevolent ruler and, except for the miners themselves, no one knew that better than the coal-company doctor.

1
King Coal

The early history of coal mining is rife with greed, inhumanity, and incredible danger. For the miner the work was one of the most hazardous of occupations, and if he (or she) survived the slag falls, the explosions, the fires and the gases, the worker was often left crippled for life. Accounts of eighteenth-, nineteenth-, and early-twentieth-century mining are haunting and sad, not only in Europe but in our own country as well.

The British Parliament appointed an investigating commission whose 1841 report on conditions in the mines, especially in Scotland, is grim indeed. Water, constantly dripping from the ceilings of the mines, stood ankle deep in the shafts and other workplaces. Because the coal seams frequently were no higher than twenty to twenty-eight inches, miners were forced to lie on their sides in the water and coal mud while they picked at the coal and struggled to load it into baskets or carts. Then they had to move their load through the water to the mine portal. It is difficult to imagine how they could work under such conditions six days a week.[1]

Intolerable as those conditions must have been for adult male miners, they were all but unendurable for the women and children. The British commission discovered not only a great number of women working in the coalpits but also children as young as five years old. They found no apparent difference in the work assigned to either sex. In fact, women were coerced to work where men would or could not go, areas so small and tight that the women had to work in mud and water up to their knees, where they were unable to stand upright. Nor were they assigned lighter loads to haul. The commission was appalled to report that women and children were treated more as beasts of burden than as human beings. No lighter workloads were accorded boys and girls; they did the same work, lifted the same weights, hauled the same quantity of coal in baskets or wheelbarrows. In addition, the commission found that girls were liable to be sent into the mines at a younger age, evidently because they were considered more tractable and cleverer than the boys.[2]

The commission noted that the proportion of boys to men in the mines was ninety to one hundred, but it made no mention of the proportion of women to men or girls to boys. Apparently the only plus for females in the mines was that they received the same wages as the males.

One mining authority of the time told of being confronted on one of his inspection trips by a woman hauling a load of coal so heavy that she could hardly keep from falling. Trembling and exhausted, she told him that she wished that the first woman ever to work in the mines had broken her back, thus freeing women from this terrible burden.

Conditions in America's early mines were not quite so revolting as those in Scotland, but as we shall see, they were bad enough. Females were not employed in mines until the 1960s, but young boys were employed there well into the twentieth century. Some of them were no more than eight to ten years old.

The original mines were simple enough. Pits were dug in the ground and the coal was brought to the surface in buckets reeled up by windlasses, in much the same manner that water is raised in buckets from dug wells. When these coalpits became too deep, water often collected in the bottom. The solution was to move along the seam and dig another pit.[3]

In the early 1800s drift mining in Europe replaced the primitive pit method. Drifts are shafts dug horizontally into the side of a ridge or mountain where coal is thought to be located. Tilting the shaft upward slightly ensured that water would drain toward the entrance. Coal was broken from the face of the seam with pickaxes, then shoveled into sacks, baskets, or wheelbarrows and hauled to the mouth of the shaft.[4] As coal was removed, heavy timbers were set up to support the mine shaft. Such shafts leading to the deep portions of the mine increased the dangers for the miners, both to their lives and to their health. Roof falls, poor ventilation, dust accumulation all took their toll.

As the main shaft penetrated deeper into the earth, where the best coal was usually found, the miners dug side shafts, called gangways; however, clearing these of the accumulating water was difficult. Crude pumps operated by horse power were employed. Also, the farther the miners dug into the coal seam the more difficult it became to bring the coal to the surface. It is not hard to imagine a miner struggling along narrow, low tunnels far beneath the surface, dragging his sack of coal or pushing a wheelbarrow to the main drift.

It was not long before someone came up with the idea of using oxen, mules, or ponies to drag sleds, at least along the main shaft, to the surface. Miners could bring their coal to the drift and dump it into the sleds. Later, crude wooden rails were laid along the drift, and wheeled carts replaced the sleds. Rails were laid down the larger gangways, bringing the carts closer to the miners. Then the miners filled the carts directly and pushed them out to the main shaft to be pulled to the surface by mule or pony.

In much the same manner, the slow, laborious method of picking at the seam to break off coal was hastened by advancing techniques of blasting.[5]

As the mines became deeper and the gangways more numerous, ventilation became a serious problem. The challenge was to suppress and flush out the deadly *damp* (carbon dioxide or carbonic acid was known as chokedamp or afterdamp, and carbon monoxide or methane was called firedamp). In the early days many a miner was asphyxiated by these often odorless and colorless gases, and many a mine explosion resulted from their accumulation. Such gases still present problems in modern mines. An old poem, "The Miner's Doom," describes the results rather graphically:

> The choke damp angel slaughters all,
>   He spares no living soul!
> He smites them with a sulphurous brand,
>   He blackens them like coal!

9

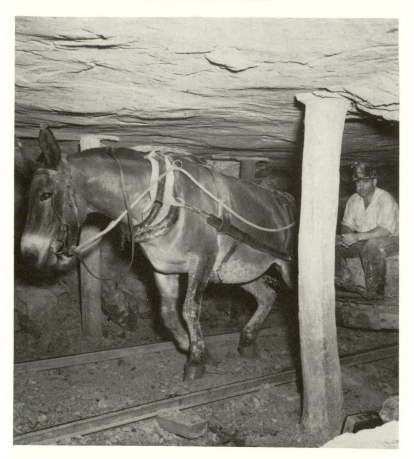

Mine haulage by mule

In early European mines this problem was solved at great cost to life and limb. When a mine was judged gaseous, the miners were called out except for one poor soul who, clad in wet leather clothes and a mask, was sent into the mine with a lighted torch. The object was to burn the gas out of the mine. Unhappily, the chances were excellent that an explosion would end the life of the chosen miner. Or else the coal itself would catch fire, consuming the timbers that supported the

10

roof and making the mine untenable until the fire could be extinguished. If it could be. In some mines this strange and frequently tragic procedure was repeated as often as three or four times daily.[6]

Perhaps because miners refused to sacrifice themselves, a better, though not greatly effective, method was adopted. Lighted lamps, called eternal lamps, were placed near the entrance of the mines and in spots where gas tended to accumulate. Their purpose was to burn off the gas before large and potentially explosive accumulations could form. This practice had its own dangers, and other methods were attempted, but none of them was very successful. Finally, mine operators realized that ventilation was necessary to pull the gases out of the mine shafts and gangways, replacing them with fresh outside air. Giant fans or blowers were installed that could push air throughout the mines from the entrances or through air channels cut to the surface. And to prevent gas from accumulating through the mine, doors were installed and then kept shut, opening only for the passage of carts or miners.

Equally explosive was the accumulation of coal dust. This became especially problematic after blasting was introduced. The solution was to broadcast a powdery combination of chemicals, consisting mainly of limestone, over the area. It replaced, to some extent, the use of water to settle coal dust.

A good deal of mining lore was transported from Europe to the mountains of Appalachia, as were a goodly number of the early mining engineers, who came from Scotland, England, and France. It was not until modern mining methods were developed, however, with a greater reliance on machinery than on humans, mules, picks, and shovels, that mining became less arduous and less hazardous. Even so, hand-loading coal into carts was standard practice as late as the 1930s.

Shortly after the Civil War, King Coal began his reign in the mountains of West Virginia, dominating both the economics and the politics of the state for many years. When the soldiers finally laid down their arms, the wild, ancient mountains of the western and southern sections of the state were relatively unpopulated. People were widely scattered in small settlements tucked into coves and the narrow river valleys. The rivers provided the chief mode of transportation to the outside world.

Mountain folk lived the lives of hardy pioneers: hunting,

11

A miner testing for gas

trapping, and fishing to supplement what they produced in gardens and in the pastures that they had cleared in the forests to raise sheep, cattle, and hogs. Occasionally, they would make the long trek to the nearest community for flour, sugar, salt, and perhaps coffee. It was a free and adventurous life but difficult and dangerous. Small wonder that the mountain people of Appalachia have earned a reputation for independence and self-reliance.

Originally, entrepreneurs were drawn to the region not by coal but by the forests of the worn mountain ranges. They started with small sawmills along the rivers. Later, following the invention of the band saw, a large, thriving lumber industry developed. This activity in turn beckoned the railroads. Spurs from the main tracks were extended up the narrow valleys to haul lumber out to markets in towns and cities in the East. Lumbering, with its promise of wages, also began to draw the mountain men out of their coves and small holdings to work either at the sawmills or on the expanding railroad. However, because the population of the region was so sparse, the railroads began recruiting their labor force among blacks fleeing the sharecropping system of the Deep South after the Civil War. They also drew workers from among the immigrants who flocked to America to escape hard and unsettled times, especially those from southern and eastern Europe

While it was well known that huge coal deposits underlay southern and western West Virginia, little mining had been done before the Civil War. Coal provided fuel for the salt industry that was situated chiefly among the Kanawha River and for the blacksmithing that was necessary for both farmers and the burgeoning railroads.

Resta Cheuvront of Murrells Inlet, South Carolina, who once lived in a coal camp at Owings, West Virginia, writes that her grandfather had a small coal bank, apparently a vein of coal, extending into the side of a hill, and that he would dig coal for his neighbors, hauling it from his pit in a wheelbarrow. She said he received ten cents a ton for his coal and would use the money to buy eggs and corn, for, as she noted, money was hard to come by in West Virginia in her grandfather's day.[7]

Some coal was shipped to market on flatboats down the Kanawha, but in general little mining was done, especially deep mining. Deep mining was not even necessary, since coal seams were exposed along the hillsides ready for the taking.

These exposed deposits in the end gave King Coal a foothold as the railroads, in extending their iron tentacles in all directions in the 1850s, began to thirst for coal. The sight of those exposed black seams above their spur tracks told the railroad officials just where they could get all they needed. Since the exposed coal was soon exhausted, miners began following the veins into the mountains, and deep mining was born.

It is said that in one early drift mine coal was rolled down the mountain from the mine entrance in large chunks. At the bottom of the 700-foot chute a large bin collected the chunks and prevented them from rolling into the river.[8]

People came upon coal in odd ways. In one place a sawmill operator started a big log down the mountainside toward his mill. The landowner watched the log uncover a rich vein of coal as it tore down the underbrush, and that was the birth of yet another mine.[9]

Thus began the saga of great riches and great tragedy for remote areas of Appalachia in Pennsylvania, Kentucky, and Virginia, as well as West Virginia. The mountain people who inhabited the coalfields suffered the tragedy while receiving little, if any, of the riches. As the railroads began to push into the valleys, land speculators followed, lured by talk of lumber and coal. These smooth operators, mostly from the North, either sweet-talked mountain people into selling the mineral rights under their land for a few cents an acre, or simply swindled them out of ownership by rigging legal channels to abrogate the titles held by those who had lived on the land for generations. Since the locals were illiterate for the most part, they were easily shorn of their rights. The educated gentlemen from the outside, and even a handful of homegrown comrades such as John C. C. Mayo, who had gone to college and studied enough geology to be well aware of what underlay the land, found it remarkably easy to part the naïve mountaineers from their property. Most of the deeds they negotiated, later called broad forms, gave the coal companies title to not only coal but also gas and metals—to whatever lay under the surface. In addition, these deeds gave the companies the right to use the land to build roads and buildings and anything else necessary to their operations. They could cut down the mountain man's trees, pile slag heaps on his pastures, turn his creeks and ponds into sewers. And on top of such deeded rights to destruction, the companies were absolved from all liability for damages that might ensue from their activities.[10]

Those who sold their mineral rights never dreamed that what was beneath the earth would destroy what belonged to them above. Those who were simply swindled out of their land lost out to vast amounts of corporate money that subsequently bought the souls of many of the Appalachian politicians of the day. With easily purchased legislatures, it was amazingly simple to obtain laws that invariably came down on the side of the big lumber and coal companies. These land syndicates were formed in the North for the sole purpose of acquiring the coalfields. Even British bankers got into the act. Robber barons of old could not have plucked the naïve mountain people any cleaner! It has been said that the British banks paid thirty-two cents an acre for mineral rights, while some land speculators paid from fifty cents to five dollars. It is estimated that today those same rights are worth around seven thousand dollars an acre or more.[11]

Should we fault these mountaineer grandfathers and great grandfathers for selling their grandsons' birthrights too cheaply? I think not, for how could they foresee that their corrupt officials would blatantly favor the absentee landlords and the coal companies, giving them precedent in decisions regarding their fields, their crops, and even their homes? How could anyone in the last quarter of the nineteenth century anticipate the destruction by strip mining that would all but flatten their beloved hills?

Once King Coal was crowned in Appalachia, he ruled absolutely. The men who went down into the mines were symbolic of the mountain tragedy. Depression, like a pall, lingers over the region even today. The feudal system that was quickly adopted in the coalfields increased and perpetuated the poverty in a land of shallow soil, steep ridges, and depleted forests. The wealth that coal brought flowed like water out of the state to northern cities to form the basis for enormous corporate fortunes. Behind, the gutted land remained incredibly poor. What little wealth remained in the area went to banks, to the businessmen in the good-sized towns that sprang up like toadstools after a summer rain, to the many an Appalachian politician and law officer who saw to it that the coal companies could do no wrong.

Miners and Medicine

The story of coal is a sordid tale with few redeeming features. Perhaps the only one is the courage of the miners and of their families on whose backs the nation prospered for many decades—for coal was the foundation of the industrial revolution in America, the black base of the country's wealth.

2
The Coal Camps

The coal camps of West Virginia and other areas of Appalachia developed naturally from the isolation of the mines and from the mine operators' desire to control the work force. There were few roads, and the means of transportation, other than the railroads that served the mines, were all but nonexistent. The mineowner had to provide housing for the miners and their families, a company store to supply their necessities, and in some cases, where mine operators were more enlightened, a school and a church. Before World War II the coal camps were an integral part of the semifeudal system instituted by the coal companies. In some ways they were like the mill towns of the North, only much tighter in structure and discipline. Miners and their families were actively discouraged from going outside the camps for any purpose, even to live outside the camp or buy goods away from the company store. In some camps miners were paid in company scrip that was redeemable at face value only within the camp. Thus compliance with the system was assured.

In probably the majority of camps, a more sophisticated system kept miners perpetually in debt to the company. While the miners were paid their wages in cash, they were paid only once a month, and since the money rarely lasted until the next payday, they often were forced to borrow from the company, which then would issue scrip for the miner's current labor and deduct the amount from his pay at the end of the month. This method successfully kept cash within the company and the miners within the camp.[1]

Some people who once lived in these company towns object to the use of the word *camp* to describe them, preferring to call them communities. Communities they surely were, but "camp" aptly describes their somewhat transient appearance, though some lasted as late as World War II.

Little more than shacks, the houses tended to be flimsy and jammed together, sometimes no more than six feet apart. Frequently they were constructed of rough boards with tar-paper roofs. Many were no larger than two rooms, with a kitchen lean-to tacked on the back. Somewhat grander were the houses with two rooms down and two above. Often they were built in tiers up the steep slopes of a coal valley where the little level ground that existed was occupied by the railroad tracks, the tipple, the mine offices, and the company store. Such company housing offered cramped quarters for the normally large miner family. The Children's Bureau of the U.S. Department of Labor, in its 1923 report, condemned them as cold, drafty, and without proper sanitation.[2]

Cold they surely must have been, for the houses where cheaply build without weather boarding and sometimes without plaster. Hence the widespread use of newspaper to cover the interior walls and to keep the wind from blowing through the cracks. The floors most often consisted of a single layer of boards over an open foundation. Frequently cold air seeped through cracks and knotholes. Dampness was a major problem because the houses were built close to the ground. The 1923 Children's Bureau report stated that a spring ran its course under one miner's shack. The report also noted that the open foundations offered refuge to animals, including chickens and pigs, and that "vermin and unhealthful odors easily entered the house." Under the house was a convenient dumping ground for junk: "an old bedstead in which children and animals were seen playing at the time of an agent's visit was stored under one house, inviting disease and fire."[3]

Fireplaces and cook stoves provided the only means of heating the camp houses. Fortunately, the companies provided unlimited coal free of charge. But that hardly made up for leaking roofs, cracked walls, and the general lack of repair, nor

A coal town

was the sooty smoke from coal-burning stoves and fireplaces much help to the hard-working housewife or for the health of the family.

Sanitation in the early coal camps was virtually nonexistent. Few houses boasted running water. A 1922 Coal Commission report noted that then only about 14 percent had water in the home and that only 3 percent had bathtubs or flush toilets.[4] Outdoor privies dotted the hillsides behind the rows of houses. Some camps pumped water from a single source to hydrants scattered through the camp, each one shared by several families. Other camps dug wells equipped with hand pumps. In some cases, fifteen to twenty families might share the same water source. Fetch and carry was the order of the day. The U.S. Children's Bureau 1923 report noted that, because the hydrant water was frequently rust-colored, many families preferred to get their water from springs, shallow wells, and even creeks, most of which were probably thoroughly contaminated. In fact, running water anywhere near the camp was far from sanitary. Chickens, pigs, children, and trash contaminated the creeks, many of which ran below the houses and their privies. In a somewhat horrified tone, the bureau report described a privy set above a house on a steep hillside, while several families took water from a spring directly below. Small wonder that for decades typhoid fever was the scourge of the mountains, while the doctors had little in the way of medicine to combat it.[5]

Rear Adm. Joel T. Boone's 1946 study, "A Medical Survey of the Bituminous Coal Industry," found conditions little better than those described by the Children's Bureau. Of 1,154 coal-camp houses, less than half had piped water, and few of those had bathrooms. Seventy-five percent of the coal camps still relied on privies. On the steep ridges, heavy rains washing in torrents behind the houses all too often caused privies to overflow, spilling human waste into the yards, the paths, under the houses, and even into the streets.[6]

In his "Memories of a Coal Camp Physician," Dr. L. R. Littleton (formerly company doctor for the Blue Diamond Coal Company at Leona Mines, West Virginia), wrote that even after World War II, when he was a camp doctor, conditions were still primitive beyond belief. He described a visit Mrs. Littleton made to one miner's home on a hot July day. The pregnant wife offered milk or coffee, but Mrs. Littleton, obviously aware of the privy out back and the general lack of sanitation, refused, stating that both milk and coffee made her sick. The miner's wife then handed her a glass of water, which Mrs. Littleton, out of excuses, could not refuse. As Doctor Littleton remarked, "In retrospect, hot coffee would have been a safer choice." He went on to relate his sister's shock when on a visit she discovered that water from her brother's company-house bathroom ran down into a small creek transversing his yard.[7]

I can remember being foolish enough to drink from a creek and being dosed with castor oil and royally spanked by my mother, who was a registered nurse and very much aware of what such creek water could contain. Fortunately, other than the castor oil and the whipping, I escaped the possible consequences of what was probably heavily contaminated water.

Doctor Littleton described some of the camp houses, which

he observed were built against the hills so that the front of the house with its porch stuck out into space, held aloft by posts. He told of a funeral party gathered to mourn the victim of a mine accident and crowded on one such porch. It gave way beneath the weight and several of the mourners "rolled down the hill."[8]

I can well remember those houses built in long rows, all alike in their shabbiness and their coating of coal dust. I can remember those porches on stilts and the privies behind, and the railroad that ran beside the rust-colored creek at the foot of the hill. In comparison, our company house was almost grand, for camp doctors rated up with the company-store manager and the mine supervisors.

Katherine Bryant of Liverpool, West Virginia, who once lived at Browns, near Kaymoor, West Virginia, wrote that she was born in what was called the New River Canyon, a valley so narrow that she said you almost had to lie flat on your back to see the sky. She described the camp as consisting of two rows of houses called "Jennie Lynns," which were built, a little deeper than they were wide, on posts, with a roofed porch in front. She mentioned that the inside walls were papered with newspapers as much to keep out the drafts that found their way through the weatherboards as to make the rooms seem clean. A light bulb hung from its cord in the ceiling of each room. Her family's house was located beside an inclined haulage track that ran loaded coal cars down from the mine portal to the tipple and hauled the empties back up to be refilled. It must have been a somewhat noisy place to live.[9]

One thing was plentiful for everyone—coal. We had all the coal we needed to heat our home and cook our meals. Children gathered the coal, for the most part, for it was almost always offered free by the company. In some camps it was even delivered to the homes, but perhaps more often, youngsters climbed aboard slowly moving coal cars to throw out what was needed, then jumped off to collect their bounty. Since most of the houses had fireplaces and grates and iron coal stoves, no doubt the consumption was high.

Katharine Bryant described the method of obtaining coal at Kaymoor Company, where they build a special short track to shunt a coal car over to a large bin built beneath the track. By opening the door in the bottom of the car, they could fill the bin, and miners' families could send their children to haul

Caples, West Virginia: A typical coal community

it home to fill the coal house, a small building in the backyard or garden. In summer when the coal was gone, the children would sweep out the coal dust and play house. Since there was a door and usually a window in these small buildings, they were ideal playhouses, especially for children who had little more than a tire swing as play equipment.[10]

Company officials, store managers, and camp doctors rated the best houses, with mine foremen rating housing slightly less grand. These better houses were the first to be improved with plumbing and electricity. Usually set apart, either on the hillsides above the huddle of miners' homes or in an across-the-tracks situation, they were sometimes handsome and imposing houses.

In the early days before World War II there was little that was green in the vicinity of the mines. Even the trees on the neighboring hillsides were cleared to provide wooden rails for the mine tracks and posts to hold up the mine roof. Sometimes there was room for a vegetable garden, a few chickens, even a pig or two behind the house; in general, however, paths, roads, and backyards were blackened by coal dust. Garbage and animal and human wastes choked the once-clear creeks. Smoking slag piles, which grew with the years, often cast an acrid pall over the valleys and sometimes stung the eyes and irritated the nose and lungs. Acid runoff from the mines killed the plentiful fish that once had been the mainstay of the mountain diet as well as an abiding pleasure for those who caught them. The early camps were not pretty, and as the Boone report pointed out, many of them never improved.

One miner's wife, Estella Akers of Hinton, West Virginia, wrote of the difficulty of keeping a clean house:

> Miner's wives were mostly dirty housekeepers because living in coal camps with dust blowing continually wasn't an easy job, so bedbugs and cinches [cockroaches?] were in most houses. Once, when we were going to have to move so he could work, I had to take a house with these creatures in it. We'd waited our turn for months for a house. . . . As soon as I went to look at it, I smelled "them." Again I cried, then I got to work.
>
> I scrubbed one room with lye water as far as I could reach on the walls and floors. Dried it and put all my furniture in the middle of the floor in lids with turpentine in them.
>
> I bought those big rolls of heavy building paper—grey background with pink flowers and green leaves. Every dishpan of paste I'd make I'd stir in one cup of turpentine. I'd almost pass out from the fumes, so you can rest assured the bedbugs did the same, only for good![11]

Martha E. Taylor of Meadow Bridge, West Virginia, had this recollection of Little Sewell Mountain:

> In the coal camps where we lived, [the houses] had a central chimney from the ground up through the second story. There were two open fireplaces with grates for burning coal downstairs and one in each of the two bedrooms upstairs. The fireplaces were in the living and dining room. The kitchen was sort of a shed lean-to. Sometimes the pipe from the kitchen stove went directly through the ceiling, and others had tile used for a flue.
>
> We were lucky, for all the places we lived in coal camps we had our own pumps on the back porch and our own backhouse, instead of, like some others, used by four or five families.

23

In the winter large blocks of ice were cut from the frozen river and stored in an icehouse. In the summer we could get a large block of ice for five or ten cents. Just think of this ice coming from a river that all sorts of wastes were dumped into—how people survived is a mystery![12]

Outside the mine itself, the company store was the focus of much of the community's activity and, as in many small American towns, the major meeting place. Like the company-owned housing, the store was necessary because of the extreme isolation of the mines. But even where mines were situated near towns, mine operators frequently required their workers to trade at the company store, the penalty for not doing so being instant dismissal. The reason? The stores turned out to be immensely profitable, charging as they did considerably more than stores in towns. It has been said that during the Great Depression the profits from the company store saw many a coal company through the worst of that economic disaster. Coal miners realized they were being had, that prices in these stores were far and away higher than what town stores offered, but the difficulties of travel, the threat of the loss of one's job, and above all, the ability to charge until payday, combined to keep them as captive customers. In addition, many miners were paid in scrip that was redeemed at face value only at that particular company's store; taking scrip elsewhere meant receiving discounted value. This too kept the miners shopping at home.

The company store was really an old-fashioned general store, stocking clothing and tools as well as food, as described by a former camp resident:

They carried everything from meat to Sunday-go-to-meeting clothes. I can still almost smell all the delightful aromas in the store mixed up with the smoke. Each morning they would mop the old wooden floor with motor oil to keep down the dust. This also added to that certain smell that only a company store could have. My favorite spot was the candy counter, with its rows of huge jars of every type of stick candy.[13]

I am in complete accord with the above (written by H. Neil Blount of Salinas, California, who once lived at Putney, the Kanawha Company's camp). The company store was my favorite spot as well. It was more than a store; it was actually a social hall, always crowded with miners, their wives, and children. There was a large porch out front that seemed never to be empty of miners shooting the breeze, whittling sticks, and smoking or chewing tobacco.

Earlier versions of the company store, some years before my own memories of the institution, employed sawdust on the floors and carried live chickens, livestock feeds, and farm implements as well as miners' tools and explosives. The store buildings were often the most imposing in the camp, and frequently both the doctor's office and the mine office shared the structure. Since the stores were generous with credit, many a miner ended a pay period with nothing more to put in his pocket than a debt slip, called an overdraft. The old mining song was so true: "Don't call me, Saint Peter, 'cause I can't go. / I owe my soul to the company store."

Using a checkoff system, mine operators deducted rent, doctor's fees, store credit, and even such things as the tools miners needed in their work underground. From one month to the next a good many miners never saw a bit of scrip, much less legal tender. No matter how long and hard they worked, they always were in debt to the coal company.

In addition to regular family housing, many camps boasted a boardinghouse for single men. Some even offered a more elegant establishment, called a clubhouse. Lillian Lilly of Huntington, West Virginia, described life in a coal-camp boardinghouse run by her grandmother near Williamson. She wrote that while both boardinghouses and clubhouses had sleeping rooms and served meals, the former were apt to be more primitive. Boarders had to do their own wash in zinc buckets and heat them on the stove. The floors were bare and bleached by daily mopping with lye water, during which, Mrs. Lilly said, the mopper had to wear shoes, something few of the young did, especially in warm weather. In contrast, the clubhouses would have rugs on the floors and their meals would be a cut above the normal boardinghouse fare, which was pinto beans, cornbread, fried potatoes, and fried apple pie. The last, Mrs. Lily said, were somewhat greasy, but delicious.[14]

Resta Cheuvront described some of the trials and tribulations of a miner's wife. At the time of the birth of their first child, her husband was working only two or three days a week, hand-loading coal for nineteen cents a ton at Owings, a mining community in Harrison County, West Virginia. They paid ten dollars a month for the rent on their company house and a dollar a month for the company doctor, and there were many nights when Mrs. Cheuvront prayed for enough food to feed her babies. They had coal to heat and cook with, but she washed clothes on a scrubboard. She would feed her children oats with canned milk diluted with water, biscuits or potatoes, and what her husband called "poor man's gravy." Sometimes they would have a little sugar, she said. In the main, their diet consisted of beans and potatoes. She gave birth to her three babies at home, and she said she never had a baby bed for any of them. They took their baths Wednesday and Saturday in an old washtub. Their bathroom was a two-holer above the house, of which she remarked, "When we had no Sears catalog, we used corncob, where we shucked corn for cornmeal to make bread.[15]

In the early days saloons were part of the coal-camp scene, and often a red-light district was included. In both, violence erupted with some frequency. Miners were a tough lot, and for some, brawling was a favorite occupation. During this time little attention was paid to bodies floating in the river or tossed into the brush along a railroad spur. Moonshine, a historic product of the mountains (but not confined there), arrived with the first settlers, giving them an excellent incentive to work hard growing abundant crops of corn. It may still do so.

As in rowdy western towns, the saloons and red-light districts retreated, though not entirely, before the civilizing influence of women and children, as families followed their men into the region.

Children apparently enjoyed a wild, free life in the coal camps. With few exceptions, those such as myself who grew up in those communities write of their memories with a great deal of affection and pleasure. Perhaps since life was difficult for adults, and parents worked long, hard hours, they had little time or energy to oversee their offspring's activities. Even though the children too were expected to do many a hard chore, such as hauling coal for the stoves and water for the wash kettle, they had plenty of time to run free. The hills and creeks provided a giant playground.

Well past the early years of our own century, that time of freedom was apt to be short for most youngsters. Boys often went into the mines at the age of twelve or thirteen, sometimes even younger. Girls often were married and had their first baby by the age of fourteen. From then on, these children were trapped in a cycle of drudgery and poverty.

The U.S. Children's Bureau reported that, even after West Virginia finally enacted a 1913 law making it illegal for boys under the age of sixteen to work underground in the mines, enforcement was lax.[16] Boys as young as ten were employed, and many a boy went into the mines after his father was killed in a mine accident. Not only did this provide the family with revenue but also it was the only way to keep a roof over the family's head. A widow without a son working in the mine was forced to move out of company housing.

Still, there were happier aspects of growing up in the tightly knit coal community. Mrs. E. Houston Halstead of Scott Depot, West Virginia, who once lived at the camp of the Solvay Colliers on Paint Creek in Fayette County, West Virginia:

> Some of the happiest days of my life were spent in the coal camp. Children in those days felt free. We roamed the hills, played in the creek, dodged snakes, hunted nuts in the fall, gathered the beautiful wild flowers in the spring and summer, gathered in a group in the evening and sat in a circle in the bottom between the company store and our house, telling ghost stories (the miners were usually very superstitious), sled riding in the snow on those hills in the winter, and even used the creek when it froze solid, played hide and seek ad infinitum. I don't recall that I ever had to ask Mama what to do to entertain myself.[17]

I have somewhat similar memories during my boyhood in the coal camps, although I shudder now at some of the daredevil things we did. The coal cars were shunted back and forth just a few feet from our yard as they were loaded under the tipple. Some of the other boys and I would climb aboard to throw coal down to be collected later. We were careful to pick loaded cars, not only because they contained the coal we wanted but also because the empties were taking their turn under the tipple to be filled with tons of coal. One boy almost lost his life when he fell into an empty car. The sides of the cars sloped inward toward a chute in the bottom, making climbing out difficult, if not impossible. Fortunately for this boy, a man witnessed his slide into the car and men at the train's next stop was notified. The boy was retrieved before coal was loaded. If his predicament had not been noticed, the

Baptism in Paint Creek in Fayette County, West Virginia

broken arm he suffered in his fall would have been the least
of his worries.

In some of the early coal camps, more enlightened mine
operators provided a school, usually no more than a single
room with rough log benches and a coal-burning stove. Little
more than reading, writing, and rudimentary arithmetic was
taught there by a teacher who dealt with all grades. Few
children went much further than the fourth grade, for as we
have noted earlier, child labor was a factor in the coalfields
well into the twentieth century. After World War I, as the
population increased in the region and towns sprang up along
the railroads and the few good roads, district schools were
established. Many a coal-camp youngster had to go to school
on a train, but the result was a far better opportunity for a
decent education.

My own early education had its gaps. My father moved us
to one coal camp where I did not attend school for several
years. One early school that I do remember was a large wooden
building with two or three grades in one room. Through the
haze of memory I recall one incident all too clearly. The school

principal disciplined a boy by throwing him against the blackboard, telling him fiercely, "I'll jerk your eyes out!" I believed firmly that he would do just that, and instantly became a model of good school deportment from then on.

Other more-enlightened coal operators build a church in their camps for the miners and their families, who tended to be intensely religious as well as superstitious. Perhaps the hazards of their occupation accounted in part for this fervor. Jim Blankenship of Davis, West Virginia, recounted memories of the religious earnestness when he lived at the Clean Eagle Camp:

> I remember the baptizings in our swimming hole. There would be two hundred people gathered up as the preacher stood waist deep in water preaching hell fire and brimstone, and people would file out to him one by one and get dunked and come out shouting and praising the Lord.
>
> My mother had many prayer meetings at our house that seemed to go on forever. I wasn't allowed in the room, and for a long time I thought religion was a very strict, hush-hush thing that was for adults.[18]

The old coal camps were communities, in every sense of the word. In many ways they were far more united in spirit and action than most communities today. When tragedy struck, as it so often did, everyone turned out to help, to mourn, to assuage the grief of the survivors as best they could. When a family got into trouble, everyone rushed to its aid. Lillian Lilly commented that, while all the neighbors "knew everything about you that you knew yourself, if there was a sickness, they just came over and helped. If you were a sot, they still helped, and if you were too lazy to work, they fed your children."[19]

It was this spirit of togetherness that caused so many former coal-camp residents to remember the good things about King Coal rather than the poverty, the back-breaking toil, and the sense of impending tragedy that hovered over their everyday lives like a cloud.

3

The Miner's Lot

The miner is a true occupational hero. Every time a miner goes underground, life and health are at risk. Even in these more enlightened times of strict federal regulation, mines are never absolutely safe. There is always the possibility of roof falls, fires, and explosions in addition to accidents involving mine machinery. Until the last several decades, mining was a daily lottery for the miner, who never knew when he entered the mine portal at the beginning of his shift whether he would emerge at the end of it. As we have noted, excellent books have been published that portray the hazards of mines and the life of miners, and we have no wish to offer competition. It is essential, however, to understand the miner's lot in order to comprehend what kind of medicine the coal-company doctor had to practice. We are talking, of course, of the period before World War II.

A 1939 report from the U.S. Bureau of Mines stressed the special circumstances that miners encountered when coal was king. It emphasized the importance of miners' health and safety and stressed the responsibility that the industry had for its employees, particularly since mining was an extremely dangerous occupation. It also noted that miners, while subject to the illnesses and possibilities for injury faced by many other industrial workers, must be especially fit physically and must possess even more than ordinary strength and health, since mining required the daily loading of six to ten, and sometimes as much as twenty, tons of coal or of the rock debris associated with extracting coal. A miner might be required to wrestle a drill weighing two hundred pounds. He might have to walk or run through the hot, muddy galleries and shafts for several miles, while following a mule pulling a loaded coal car down the low-beamed workings, or even just to reach a work station. The task of getting to and from one's work station each day often included climbing up and down ladders several hundred feet high, or even several thousand, in some cases.[1]

Miners whose health failed, the Bureau of Mines commented, were assigned to the scrap heap, and many a former miner was forced to turn to a less-arduous occupation such as raising chickens, managing a bar or poolroom, or even becoming a watchman at the mine they once had worked in. The one bright spot in all this, the bureau seemed to say, was that a miner's hard life might protect his health in the long run more than a sedentary occupation would. I am not at all sure miners would agree.[2]

Certainly, for the early-day miners before about World War I, there were few advantages and a host of disadvantages. Well into the twentieth century mining was a pick-shovel-and-wheelbarrow operation. Carrying wooden rails into the mines at the beginning of their shifts, the miners themselves laid down the tracks to enable the mine cars to move forward in the shafts close to the face of the coal. When the cars were full, mules or ponies would haul them to the portal. To loosen the solid coal veins, the miners would undercut them with their picks, then use an auger attached to a breastplate to bore a hole in the face. This they would fill with black powder and a core of dust, preferably clay, with a wick extended. Once the blasting powder was tamped in carefully, ignition was a simple matter of lighting the wick, then retreating as quickly as possible. Since some shafts were no more than three feet high, forcing the miner to work on his knees or even lie on his side in the mud, and often in water, the retreat from shooting the face must have been difficult. After a few years of such contortions in restricted work areas, arthritis would set its firm grip on a miner's stressed joints. Describing her father, a miner of forty-four years, Hazel Martin of Oak Hill, West Virginia, said that, when she was a child at Rock Lick: "Many

Boys with mules

times his pant legs was frozen because he had to work in water. When he came out of the mines the cold froze his pant legs up to the top of the thigh. He was bothered with what he called lumbago. He went to work when he was sick. Mother made thick pads and sewed [them] in my father's long underwear to keep his back from being raw from hitting the roof."[3] It is hard to imagine how a person could work under such conditions, especially when he had to swing a pick or shovel coal into a car. It is not difficult, however, to comprehend what a few years of this could do to a man's body.

In earlier times miners went into the mines before daybreak and came out after sunset, working twelve to fourteen hours a day, every day of the week except Sunday. Men, mules, and ponies rarely ever saw the sun. Mules and ponies that were stabled underground frequently went blind. Bonnie Prather of Charleston, West Virginia, wrote of when she was a child at Snaketown:

> I remember my parents retired to bed very early in the evenings, usually around 7:00 or 7:30 P.M. because they arose for a new day about 4:00 A.M. My dad left for work around 5:00 A.M. after a very

hefty breakfast. He carried a metal lunch bucket that held water in the bottom and food on top. [A] carbide lamp was fit into the front of his hard hat. His "bank clothes," as they were referred to, always seemed to be dirty with coal dust, and I don't remember that they were washed often. They were put in the corner and put right back on the next morning. Dinner was always on the table when he came in, which was about 4:00 P.M.[4]

Madeline Craddock of Saint Albans, West Virginia, wrote of her early life at Lens Creek: "My best memory of my dad's work in mines [is that] he had to leave home at 3:00 A.M., walk through [the] mountain[s] to Davis Creek, West Virginia, to work, walk back home, and in the summer he would rest awhile, eat his supper, get his hoe on his back, go back on the hill to the cornfield and work until dark, six days a week (five miles to work, ten miles a day)."[5]

Sanitation in the mines presented a nasty problem, especially since the men stayed such long hours underground. In the early days facilities were nonexistent. Miners relieved themselves at any convenient spot in the workings. Excreta and litter accumulated, and rats frequently found a haven in the muck, made worse by stagnant water black with coal dust. The smell must have been overpowering, given the lack of proper ventilation in those early days. Oddly enough, miners welcomed the presence of the rats because the rodents could sense movements in the mountain above the mine roof and would make a quick exit, warning miners to do the same. Thus, with their ability to sense rockfalls that were still undetectable by humans, they served much the same role as the canaries and white mice that coal miners carried into the mines to warn them by their demise that deadly odorless methane or carbon-monoxide gasses were present.

Survival for the miner and his family depended on how many cars he could fill with coal each shift in spite of those gruesome conditions. Once the coal face was blown and the coal shattered into manageable chunks, the miner shoveled it by hand into the car, or if the working area was too cramped to allow for that, into a wheelbarrow, which then had to be pushed back to a roomier area and reloaded into a car. Each miner was issued brass discs stamped with an identifying number. He would hang a disc on each car that he filled as his claim for pay. In some mines where miners distrusted each other, he might have to hide his disc in the coal in order to prevent someone else from removing it and replacing it with their own designated number down the line.[6]

In his memoirs, Howard Ross, Sr., described mining as a "back breaking business" that "exacted a deadly toll in lives and limbs, not to mention broken health."[7] His father, a miner, described the early kerosene lamps that were later replaced by carbide lamps, which in turn made way for the far-safer battery-cap lamp. In the 1800s miners such as Ross's father would drill the coal face with augers, then light the escaping gas with their kerosene or carbide lamps, and "a flare would shoot out ten or twelve feet." It was to be hoped that there was no critical accumulation of coal dust or gas. Once they had disposed of the escaping jet of gas, they would blow the face with dynamite, then pick-and-shovel it into the cars.[8]

Howard Ross was born and raised in the coalfields. He recalled one tragic incident which occurred when he was a boy. A miner, evidently unable to provide for his family, decided to commit suicide, and in his despair bought a bottle of carbolic acid: "Coming back down the road, he turned the bottle up and drank it. He died with his hands wrapped around a strand of barbed wire along the road, and his screams will always be in my memory."[9]

Earl D. Henderson is a retired miner living in Robinette, West Virginia. When he was in his eighties, he recalled in a lively letter his own youthful experiences in the mines:

I ran away from home at the age of fifteen in the year of 1919. . . . I came to a small coal-mining camp in McDowell County that was called "Lone Jack." . . . I went to the mine foreman and asked for a job and, as they were needing men, he asked me if I'd ever worked in a coal mine. I lied to him like a yellow dog and told him that I had. I don't think he believed me, but he told me to come up on the hill the next morning to work.

So next morning, bright and early, I climbed that steep hill to what was called "the bottom team," and stood ready, I thought, to get into the man trip, three small cars pulled by a mule.

Well, that bank boss walked up to me and said, "Son, where is your lamp?" (They used carbide lamps on a cap on your head.) I had on a big, broad-brimmed cowboy hat. I said I didn't know I would need a lamp, that the mine foreman had told me I would be on the day shift.

You talk about a gang of coal miners howling and laughing, that

Mine haulage by rail car

crowd sure did make me feel little! I almost wished I was back home off-bearing at Grandaddy's sawmill. But that dear old bank boss went into the shop where his office was and located a lamp and a cap to loan me. And I climbed into that car and sat straight up, ready once more to go into that old coal mine and dig coal (I thought).

But once again I was wrong. Someone told me I'd have to lay down in that car, for the top was only about four feet on the main entry and only thirty-eight inches where we loaded coal. So I stretched out, and the mule driver cracked his whip, and we started underground, it seemed to me at about forty miles an hour. (I guess it was about five miles an hour.)

As I lay there in that car, every story I had ever heard of coal-

35

mine explosions, roof falls, cave-ins, and flooded mines flashed into my memory, and I began wishing that my address was Earl Henderson, Route 2, Saltville, Virginia. . . . And I wondered where they would send my body, if they ever found it, for I had given my home address at Minot, North Dakota, a place I'd never been or hardly heard of. I just happened to see that name in a newspaper on the train the day I come to Lone Jack. Well, we finally got inside, and after bumping my head several times, I began to learn to either stoop or get down and crawl.

They called the crew I was with the "Green Bunch." We had an old, experienced miner with us to shoot the coal and see that we set the timbers and to try to keep us from getting killed or killing each other. He gave us a big coal scoop, or shovel, and started us loading those little cars, and he started boring holes in what I called the front wall; they called it the face. He bored those holes with an old seven-foot crank auger, having a breastplate which he put around his belly instead of his breast. They shot coal with black powder, made up themselves in long cartridges which they slipped back into the holes; then they stuck what was called a needle of copper into the hole and into the powder; then they tamped up the hole with coal dust with the needle still in there. Then the needle was withdrawn, and they put in a squib, a small fire crackerlike thing with a fuse of sulphur that was lit, and it would jump back into the hole and put off the black powder, and the coal would come tumbling down for us to load.

I was getting five dollars a day for about nine hours work, which I felt would soon make me a rich man, for the most I had every made before was one dollar a day at a sawmill. I paid one dollar a day for board at the boarding house . . . and I thought I would soon be a millionaire, but soon learned better. . . .

The camp [at Lone Jack] was made up of about twenty-five houses, with the company store and offices in the same building. The houses were mostly shacks, not even painted, with water coming from three wells with hand pumps. We bathed in washtubs, in water carried from the pumps and heated on coal-burning stoves. . . .

The people in those old-time coal camps were real people; in fact, they were folks, folks ever ready to help each other or even strangers. The place I found to board took me in, a strange boy, dead broke, and fed me and kept me until I got a job; then only did they start drawing that dollar scrip on me each day for my board. . . .[10]

Today most people would be shocked at the idea of a fifteen-year-old boy going to work in the mines or, for that matter, in any other heavy industry. But, as mentioned previously, child labor with all its attendant horrors was normal well into our

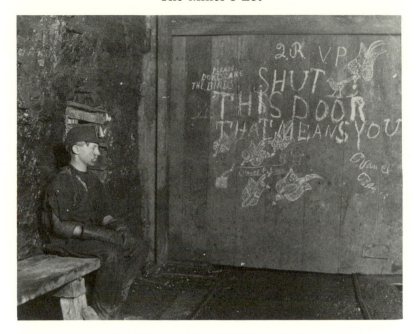

A trapper boy

own century. Lucille Radford, of Kimberly, West Virginia, related her father's introduction to coal mining: "He came from Austinsville, Virginia, at the age of thirteen and went to work in the mines at Powellton and Elkridge. Older men carried his dinner pail, as it would drag on the ground he was so small."[11]

Many of those youngsters were employed as "trapper boys." These were almost always very young—under sixteen—and sometimes were not even ten years old. They spent their long day in darkness opening and closing trapdoors for the draft animals as they hauled the loaded cars out of the mine and returned with the empties. Others worked as "breaker boys," so called because they straddled the chutes that carried the coal from the mine portal and, with heavy hammers, broke up the larger chunks of coal.[12] It was a dirty job that blackened their skin and filled their lungs with coal dust. There can be little doubt that many of these boys fell victim to respiratory problems, even black lung, in their later years.

In 1915, West Virginia amended its child-labor law to make it illegal for children under fourteen to work in factories and under sixteen to work in mines. Unfortunately, as is often the case with good laws, enforcement was inadequate, and boys of fourteen still worked in the mines, especially during the summer months when school was out. Like their elders, the young suffered injuries and damage to their health, often without compensation of any kind from the company.

In her 1923 report, Nettie P. McGill of the U.S. Children's Bureau stated that of fifty-two boys who worked full-time in the mines, ten had been injured, with one injured twice. Four of those youngsters were under the age of fifteen; two were in the mine illegally. Their injuries ran the gamut of split fingers, lacerated or burned legs, sprained backs, injured knees, and broken bones, causing disabilities lasting from one to seven weeks. Families of the boys claimed that only four of the ten disabled received any kind of compensation from the coal company. One boy, fourteen years of age and working as a miner's loader, suffered an injured back in a slate fall and was incapacitated for six weeks. He received no compensation from the company, perhaps because he was under age. Under the law, however, if a worker was deemed disabled for at least eight days, he was entitled to compensation for his injury.

In another case, the Children's Bureau found that the company did not compensate a sixteen-year-old trip rider who broke his leg when he was thrown from a mine cart. It seems that in the cases of child employees, especially those working illegally, the coal companies simply ignored both child-labor and worker's compensation laws.

Even those children who received compensation were not exactly royally rewarded for their pain and suffering. Of the four boys in McGill's 1923 report, the company paid $7.98 to a fifteen-year-old coupler, who had been run over by a motor and incapacitated for a month, and $25.00 to a sixteen-year-old, also a coupler, whose leg had been badly burned.[13]

Children and their elders were exploited in the coalfields long after laws were on the books designed to protect them. If one considers that 1 out of 225 miners could expect to die in the mines, employment of children seems especially callous. In truth, the history of child labor during the industrial revolution is a shameful chapter that did not completely end in this country until as recently as World War II. Such practices continue today in much of the Third World and among illegal aliens in big cities in the United States. Until the Second World War both the individual states and the federal government were incredibly slow to protect workers. The unions eventually provided, at least in part, that protection, but during the depression they were still struggling to gain a solid foothold in the mines and in industry.

The first mine safety law evolved as early at 1870 in Pennsylvania after a grim mine fire at Avondale trapped and killed 179 miners.[14] By 1900 as many as twenty-one states had mine safety laws of one kind or another on their books, but enforcement was lax, and miners continued to be injured and to die needlessly. West Virginia had passed a mine safety law in 1879, but it proved to be of little or no help because mine inspectors were few and far between. It was hinted that they also were easily bribed to ignore infractions.

But the time the United Mine Workers' Union was formed in 1890 the struggle for better and safer working conditions in the mines had intensified. The history of that struggle in the coalfields was especially bloody. In the end the union was strong enough to pressure the federal government to pass and enforce regulations to protect the lives and health of miners, but only after a costly national strike in 1946, called by John L. Lewis, the famed head of the United Mine Workers of America, when the coal companies turned down a union request for a welfare package in the new contract. President Harry S. Truman invoked the national interest to seize the mines and continue operations under government direction. Under the federal direction a health-care plan was initiated, financed by the government, which was the precedent for the plan adopted when the strike ended and a new contract was negotiated between the union and the coal companies. The union won two concessions: a welfare and retirement plan funded by royalties on every ton of coal dug, and a government study of the health care that miners and their families were receiving. The study revealed that the health care in mining communities was similar to that reported in the Boone study for 1922, with very little change and many inadequacies, especially in the care of miners disabled by mine accidents or by special health hazards such as black lung and arthritis. One of the first tasks of the UMWA Welfare and Retirement Fund was to identify and provide medical and rehabilitative care for

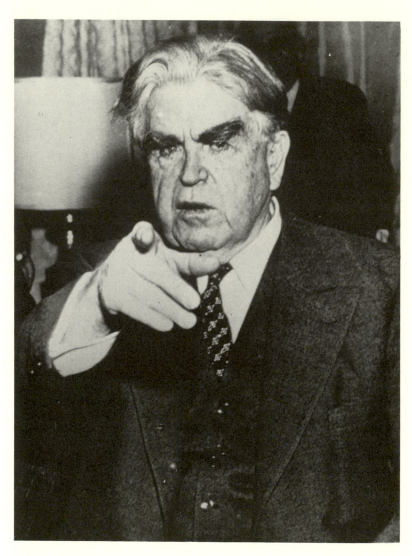

John L. Lewis

thousands of disabled miners, some of whom had not received medical help for several decades.[15]

That original health-care system has altered with the years to address new problems, and it is generally agreed that the health of miners and of their families has greatly improved. Unfortunately, stronger safety laws and better enforcement of them had to wait until after the deaths of 119 miners in 1951 at West Frankfort, Illinois, and of 78 miners in the Farmington explosion of 1968, described in chapter 10 below.

The long, ugly battle to unionize the coalfields is beyond the scope of this book, but O. E. Palmer of Cedar Grove, West Virginia, has contributed an interesting account of the violence associated with the drawn-out conflict. One of his memories was as follows:

> In the Tomkins homeplace, where one railroad ran behind the house and one in front of it. One faction, behind the house, since it was elevated above the railroad, and one faction in front of the house, since it also provided [a] depression, had a gun battle between themselves firing around the house. One of the men in the front section was shot, and he crawled up on the front porch, and my grandfather dragged him into the house, where he died.
>
> Another time around 1927 during a bitter UMWA strike, Valley Camp Coal Company brought in two railroad cars of strikebreakers by rail, and it was manned with machine guns on top and at each end, and as they went up the Kelley's Creek, the strikers fired on them with rifles from the hills and in turn they fired back.
>
> [One of the] two of the most notorious coal "kingdoms" [was] Carbon Fuel Company on Cabin Creek, where you could only enter by train (their train), and as you passed into the coal camp, you passed between two brick, concrete, and steel guard towers with guards and guns. Then you stayed therein, paid rent, paid utilities and all connected company charges, whether you had the same or not.[16]

Palmer knew a man working there who found that his pay statement mistakenly had a charge on it for water. He asked the front office about it, protesting that his house had no water, not even a well. He got his water from a spring in the hollow behind his house. They informed him they owned the spring, and they would continue to charge him for water.

On a note of sarcasm, Palmer remarked:

> [The] Crown Hill was a little better by the time I went there in 1930. All they did was shoot at the C & O trains as they passed

Coal-company guards

through town, make and sell moonshine in the hills, bring the product through an old mine into town. If my memory serves me right, they only killed six while I was there.

Being "company," I worked the . . . strike in the 1930s. . . . When the supply of coal became critical, we ordered in six coal cars [to haul coal out]. Six of the strikers, rifles in hand, came out at the start of the belt line and told me that if I loaded the coal they would kill me. I pointed out that the coal belonged to the company, had been mined by the men facing me, and that they had mined and been paid for that coal under a valid union contract. I loaded the six cars, and since you can see that it is now being related by me, it would seem that they backed off.[17]

Matters were seldom settled so easily during the brutal struggle, which, if one considers the history of coal mining and the nature of the miners' lot, ended finally in triumph for the men who dug under the mountains.

4

Death and Danger in the Mountains

Until the 1930s little if any attention was paid to the prevention of mine accidents. When a miner was killed or disabled, he was simply replaced. Miners often said—and believed— that the coal company operators would rather lose a miner than one of the mules that pulled the coal cars. Safety regulations in West Virginia mines were as lax and as little enforced as the child-labor laws. The state had the unenviable reputation of being the mining state least concerned about the safety and welfare of its workers. One West Virginia governor allegedly remarked that it was entirely natural that miners should be injured and killed in the course of their labors. The coal companies literally got away with murder well into the twentieth century.

Over time, roof falls killed more miners than the more-spectacular explosions and fires. In their efforts to prevent such falls, miners denuded the forests around the mines to obtain the timbers that they used to support the mine shafts. Even so, it was not until the technique of bolting was employed that casualties due to roof falls decreased.[1]

Bolting consisted of driving heavy iron bolts with wide iron plates attached up into the rock. In some mines, timbering and bolting were employed in tandem. Some miners preferred the old method of timbering because they felt that it gave

43

them better warning of rockfalls, since as the rock shifted and cracked, the timbers would pop and crack in turn, warning the miners to retreat. In mines that were inclined to be gassy, bolting was not employed; when the rock strained the metal bolts, sparks were apt to fly that might ignite the gasses.

Miners have described rockfalls as the mountain coming down, and the noise as thunder. They also say that the compression of air in the shafts when a fall occurs could knock a person down.

Before federal regulations halted the practice, giant rockfalls would sometimes happen when miners began pulling down the large columns of coal purposely left to prevent cave-ins. When a mine was played out, the coal companies would have the miners salvage this remaining coal. It was a dangerous business, and many a miner was injured or killed in the falls that resulted. One miner said of a fellow that he was caught "just like if you took a worm and slammed your foot on it."

Large rockfalls within the mines often produced "mine breaks" in the ground above. Sometimes these breaks could be as deep as one hundred feet.

Death and injuries occurred because of accidents from other causes, including runaway mine cars, the electric cables (when mines began to use more modern machinery), and of course, the ever-present danger of fire and dust or gas explosions. Coal dust and the buildup of odorless methane and carbon monoxide are ever-present dangers in the mines. To control the dust, rock dusting was used, or water was continually sprayed to keep the dust wet. Large ventilation fans helped to clear out methane as well as improve the air for the miners, but according to one engineer, the fans also dried and stirred up the potentially explosive dust.[2] It was because methane and carbon monoxide are odorless and colorless that canaries or white mice were used to detect its presence, miners knew that when the small animals expired they themselves were in danger. Overexposure to coal dust and mine gasses other than methane could leave a miner with lifelong health problems such as black lung, described in chapter 10.

Perhaps one of the worst mine disasters in this country occurred in 1907 at Monongah, West Virginia. It left 361 miners dead, so many that the bodies had to be lined up along the

Street scene after the Monongah, West Virginia, mine disaster

town's main street due to lack of space. They lay in rough pine coffins awaiting identification by relatives and friends.

Langdon Smith, writing in the *New York American* in 1907, gave a vivid account of what happened to two brothers and their father when the mine blew up. Perhaps their first indication of trouble was when the lamps began to flicker. Apparently they had stopped work to try to fix them when suddenly the earth shook and there was a thunderous roar. The explosion created hurricane like winds that swept through the mine shafts and galleries. The force of the air was so powerful that it blew the three miners off their feet and sent them tumbling down the shaft. The father was momentarily knocked unconscious, and the two brothers, remembering an air hole, carried him toward it. They knew they had to leave the mine as quickly as possible if they were to survive. By the time they had reached the air hole, the father had recovered, and all three began to climb toward the surface and safety. Just as they were almost to the top, a second explosion occurred, blowing them right out of the hole and clear of the mine. One of the

45

brothers was quoted as saying that no one else would be found alive, that the mine was cursed.

Rescuers did in fact find one other survivor, a Dutch miner sitting on the body of his dead brother. It was one of the tragedies of early mining practice that all the male members of a family might go into the same mine; thus, when an explosion occurred, it could devastate an entire family.

One of the rescuers who went into the Monongah mine following the explosions had an equally harrowing tale. They had to shore and brace the timbering, he said, as they moved slowly into the mine. He described how the dead lay all around him, blown into heaps by the force of the explosions and the fierce winds. He says that limbs, and even shoes with the feet still inside, were lying about.[3]

Another eyewitness described the scene outside the mine as the explosion occurred. As the earth shifted beneath the town, buildings shook and some collapsed. The thunder of the blast reverberating against the hills was deafening. Pavement heaved and cracked as the earth moved, and it was said that people and horses in the streets were knocked down. Streetcars were even derailed. Fire broke out in some sections of the town.

While all that was frightening enough, the knowledge that hundreds of miners were trapped underground sent people rushing to the mine. Families of miners on that shift and many others gathered at the mine entrance, now blocked by debris. The bodies of two miners had been blown clear of the mine and lay outside. Timbers that had supported the roof had been blown right across the nearby river, and even as horrified people gathered in front of the obstructed shafts, gas was still escaping. One distraught women who ran towards the blocked entrance was knocked flat by the force of the gas, while a miner who rushed to her rescue was almost overcome.

The rescuers faced an all-but-impossible task. Galleries had collapsed on the workers, and main shafts were choked with tons of earth where the bracing timbers had given way. Gas still escaped, threatening them with suffocation. It was slow, difficult work as they moved step by step into the mine, risking their own lives trying to reach the trapped men inside. But as they pushed ever deeper, they found only bodies, some badly mangled by the rock that had buried them. In the end, there were few survivors.[4]

Ahlena L. Wells of Huntington, West Virginia, who lived at Sturrett coal camp, observed: "My dad used to ask me if I was trapped in a mine with my mule or pony, what would I do first? Of course I knew. I'd heard him say often enough; one would kill the mule because he breathes so much more air than a man."[5]

As late as 1927 a mine explosion took ninety-seven lives at Everettsville, West Virginia. It was claimed that the explosion was due to the refusal of absentee mineowners to provide for rock dusting in the mine. Mine safety improved at a snail's pace in the state. Between 1900 and 1980, 19,826 miners died in mine accidents. Over 11,000 of those fatalities occurred before 1930. After that fatal mine accidents slowly decreased.[6]

As a child at Carbondale coal camp, Alice May of Clifftop, West Virginia, learned of her father's death in the mine the hard way:

> When I was nine years old and in . . . the third grade, I was on my way to school . . . when one of the men was coming in from work and he stopped and told me there was no need for me to go on to school because my dad was killed during the early morning hours. That was kind of a bad way to tell a kid. Anyway I went back home. This was February 3, 1949, that my dad was killed (age fifty-two years). They say that he stepped out from behind a curtain and the cars ran over him. My brother James . . . was with Dad when he was killed.
>
> Well, anyway, mom was left a widow with eight kids still at home.[7]

Mrs. May goes on to say that her mother remarried, but that her stepfather also was killed in a mine accident six years later. In the days before Social Security and the union, when a miner was killed his widow and children were out of luck. Often the company would evict them from the company house to make room for a new miner and his family, and except for what might be collected from her husband's friends, the widow had nothing. Neither did she usually have any means of earning a living for herself and her children. Her only recourse was to remarry as quickly as she could. Nor was it rare, as in the case of Alice May's mother, for a woman to lose several husbands to mine accidents.

By 1910 almost every state had some kind of workmen's compensation law on the books, but few workers injured on

Evicted miners

the job ever received compensation. It was even rarer for the surviving family to be compensated when the miner was killed. West Virginia had a workers' compensation law of sorts, but as the following case from my father's files illustrates, it did not do much for the widow or her children.

The miner died in 1927 when his head was crushed by some coal that fell from a seam face and pinned him against his cutting machine. He had been receiving ten cents a ton for the coal he sent out of the mine, working six days a week for a monthly wage of about three hundred dollars. At the time of his death he was twenty-eight years old and the father of three children under sixteen years of age, with a fourth on the way. He had worked for the same coal company for ten years. His family of five received compensation of about fifty dollars a month, which ceased when the widow remarried, a step she had to take to feed herself and her children.

A miner named Samuel Green related how a friend of his

A mine first-aid station

was so badly injured in a mining accident that he could not be moved. While another man went for the doctor, Green stayed with his friend. Soon rats began to gather around the injured man, and Green had to beat them off. Yet Green added that the miners so much liked seeing the rats that they would share some of their lunch with them.[8]

Richard E. Arneach, Sr., of Sylva, North Carolina, traveled through the camps as a trader. He described how injured miners were carried out of the mine:

The mules were often used to drag the stretchers used to carry the wounded on. The stretchers were often just two poles with a tough blanket stretched between them, and on the poles were hooks where they hooked the stretcher to the harness of the mule. If a mule could not carry the stretcher, two men could, by getting

49

on either side and lifting it up to carry it. When accidents happened in the tunnel, they often used the mule to carry the stretcher until the tunnel was large enough for the men to stand straight up.[9]

What seems a clear violation of mining safety laws led to an explosion in the Parral mine near Fayetteville, West Virginia, that killed twenty-three miners in 1906. The law in question required two outlets to a mine employing twenty or more workers, but Parral had only one opening. Even though the opening was partitioned to allow ventilation, it evidently did not provide enough air exchange to keep gas from accumulating. When the gas ignited, it fired accumulated dust, bringing down about a hundred feet of rock and releasing more gas, which in turn was ignited by the open light of a miner shifting cars in a driftway.[10]

The state's chief mine inspector urged a grand jury to indict mine authorities for violating the safety laws. The jury did indict the superintendent of the Parral mine for voluntary manslaughter, but the indictments were never pursued. In the heyday of coal-company power in West Virginia this was not an isolated instance.

Ironically, a further tragedy occurred about a year later in the nearby Stuart mine at Lockgelly as its management was trying to comply with the two-outlet law. As at Parral, the company provided only one outlet. The plan was to connect the two mines with a mile-long tunnel so that both would be in compliance. The tunnelers were within a thousand feet of their goal when an accident occurred in the shaft. A loaded car being hoisted to the top shifted and tore up a section of the shaft timbers. Since it would take some time to repair the damage, the miners were dismissed. They had started to move back when the Stuart exploded. Apparently accumulated gas had ignited, which, fed by coal dust, produced a blast that swept through the mine. It was also believed that dynamite carried by the miners blew up. The power of the explosion blew the harnesses off the mules. Eighty-five miners were killed.[11] Among the dead was a thirteen-year-old trapper boy, according to Ahlena Wells.

Not all mine accidents were the fault of the coal companies. Miners themselves, through carelessness or disregard for safety regulations, touched off more than one explosion. This was apparently the case in the Red Ash explosion of 1900 that killed forty-six miners. Although known as a gas-producing mine, the Red Ash in the New River Gorge had a good safety record for the coalfields of the time. On this particular March day, however, the fire boss was late to work, and rather than follow the custom of waiting in the main heading until he had conducted his gas inspection, the miners had already entered the mine by the time he arrived. According to the law, no miner was to enter any mine known to generate fire damp until the fire boss had inspected the mine and found it safe. The investigation after the tragedy faulted both the miners who had not waited for the usual inspection and those of the previous shift who had left a door open, allowing the lethal build-up of gas.[12]

While explosions and mine fires were multiple killers, slate falls (roof cave-ins in working areas) killed and injured more miners in the long run. The railroads too contributed their share of tragedies. A work train carrying about 350 miners blew up near Powellton in the Kanawha District in 1934, killing eighteen men and injuring about forty-seven others. The explosion was so powerful that a section of the engine cab was blown across a creek into a house, almost crushing a woman and two children asleep in their beds. Another part of the locomotive blew right through a house. Unfortunately, the miners in the first coach had gathered at the front of the car to soak up the warmth from the firebox, and the blast blew a good part of the engine through the coach, crushing them, while escaping steam burned others.[13]

Shirley Donnelly, writing for the *Montgomery (W. Va) Herald* in 1982, unearthed a list of mine casualties in Fayette County for the year 1927. Fifty-six men died in mine accidents that year, and the list is typical of the variety of fatal accidents that occurred in the mines. One man, twenty-three years old, fell twenty-four feet from a platform, which then fell on top of him. Another, nineteen years, was killed in a collision of two haulage motors. One twenty-year-old miner with two children was killed when ice fell down a shaft, smashing both the cage and the young father's skull. Some of the fatalities were bizarre. One young miner died when a mule kicked him, causing him to fall off a car, which the mule then pulled over him. Another miner left a wife and four children when he was rolled between the top of a loaded coal car and the roof of the gallery. Some thirty-six miners were killed by slate falls, which were far and away the leading cause of mine deaths that year. Several miners were electrocuted, and several more died when run over by mine locomotives or coal cars. The ages of those killed ranged from seventeen to fifty-six.[14]

In another railroad accident a landslide swept a small, narrow-gauge locomotive and its two cars and caboose off the mountainside and a thousand feet down into the valley, burying the train beneath mud and boulders and broken trees. It took days to dig out the dead from the wreckage. It was said that one survivor had his clothes ripped from his body.[15]

Some of the hazards miners and members of their families faced almost daily seem like scenes from an adventure movie. For example, the town of Kaymoor, located along the New River in southwestern West Virginia, was divided into Kaymoor Bottom and Kaymoor Top. A thousand feet of steep mountainside divided the two parts of the town. Kaymoor Bottom's only connection with the outside world was the Chesapeake and Ohio Railroad, while Kaymoor Top possessed a road. People in the Bottom, including high school students, miners, and housewives, rode a wooden car up haulage tracks to attend work, school, or visit the stores. Primary schoolteachers, off-shift miners, and others returning to their homes rode the train down to the Bottom. The first several rides must have been a thrilling experience.[16]

As late as the early 1930s children at Gauley Bridge had to take a ferry or rowboat, then a train, to get to high school. When the river froze they simply walked across the ice. Even the train ride could be an exciting experience, for on occasion some children rode the cowcatcher. The schools themselves were apt to be flimsy board buildings. One teacher complained that sheep seeking shelter under her one-room structure used to shake the whole building when they scratched.[17]

Swinging bridges, often one plank in width, spanned many a creek and river in the coal region. Crossing them must have been doubly exciting when the water was up, transforming even the mildest of creeks into a roaring flood. For example, the New River divided two coal camps that were connected only by a 500-foot-long, six-foot-wide swinging bridge about 75 feet above the river. Since the C & O railroad ran on one side of the river, inhabitants, including schoolchildren, had to make the crossing for such necessities as schooling, mail, and groceries, or to ride the train out of the valley.[18]

A New River Canyon mine location

Miners and Medicine

Yet perhaps the greatest hazards faced by the miners and their families were those common in other rural areas of the United States in the first half of the twentieth century: diseases such as typhoid, tuberculosis, and dysentery. Because sanitation in coal camps left a great deal to be desired, and large families tended to overcrowd very small homes, germs spread easily. Epidemics of typhoid, scarlet fever, and diphtheria thinned the ranks of children, and those who survived the original onslaughts became vulnerable to lesser infections. Fortunately, by World War I the mines, the lumber industry, and the railroads finally had brought the doctors into the hills.

5

Old-time Medicine

The history of medicine, from colonial times until well into the twentieth century, is somewhat dismaying. To survive, a patient had to possess a strong constitution, a good deal of luck, or both, because the medicines favored by physicians and the treatments they prescribed frequently did more harm than good.

Initially, medicine was practiced by just about anyone who wished to call himself a doctor. Since very few trained physicians emigrated to the colonies, the gap was filled by self-proclaimed doctors or by apprentices who had learned their trade by serving a few months under an established physician. The latter generally had been an apprentice himself, but had moved up a notch in skill by the trial and error of experience. Often medical practice was simply a side occupation. Many a clergyman or planter treated human patients just as they treated their sick and injured animals. Although some colleges had the study of medicine as a regular part of their curriculum, the courses served just to encourage those interested in the subject to go out and hang up their shingle.[1]

Basic medical doctrine in those days was itself rather simple: disease was caused by morbid humors of the blood and bile; doctors simply bled or purged or sweated the patient, or all three.[2] It does seem remarkable that anyone survived such treatment on top of an illness. Castor oil and calomel (mercurous chloride) were the early drugs of choice, with the later additions of opium (laudanum and morphine) and alcohol. The most dreaded diseases came and went in epidemic waves: yellow fever, cholera, smallpox, diphtheria, and measles. Malaria, tuberculosis, pleurisy, pneumonia, and dysentery added to early Americans' woes well into the 1900s. Mortality among children was extremely high, and life generally was relatively short. Medicine could not do very much to improve health and happiness, and although medical knowledge was slowly accumulating, spectacular advances were few and far between.

One of those advances occurred in Europe when Louis Pasteur and Robert Koch, among others, developed germ theory in the late nineteenth century by isolating and cultivating the bacteria responsible for some animal and human diseases such as anthrax and rabies. They demonstrated that immunity to such diseases could be developed by inoculation of an attenuated form of the disease germs, a process known as vaccination. The term vaccination was first applied to Edward Jenner's earlier discovery that inoculating humans with a small amount of pus from cowpox vesicles in cattle could protect them from smallpox, one of the most murderous of human diseases.[3]

Another significant advance was made by Joseph Lister, an English surgeon, in 1865. Subscribing to the germ theory amidst prevalent postsurgical and postpartum infections such as blood poisoning, puerperal fever, tetanus, and gangrene, Lister demonstrated that such horrors could be avoided by sterilizing surgical instruments, dressings, the operating table, and the gown and hands of the surgeon himself. Originally working with carbolic acid, he advocated the use of antiseptics and the importance of absolute cleanliness to prevent infection of wounds and during childbirth. This was a huge step forward at a time when surgeons approached their patients in frock coats and washed their hands after surgery rather than before. Unfortunately, since antiseptic and aseptic measures (such as sterilization of the operating field) tended to be burdensome and, like almost everything, not 100 percent reliable, they were slow to be adopted.[4]

A third big advance that transformed the art of surgery was the discovery that ether and chloroform could render a patient unconscious during operations, thus saving the patient grim agony and the surgeon a good bit of frustration as he tried to deal with a person writhing in pain.[5]

Medical training was initially all but nonexistent in America, although by the early 1800s medical societies were struggling to straighten out the chaos and determine just who was and who was not a doctor. The societies did manage finally to differentiate doctors who had obtained degrees from the medical schools in some of the larger cities from the apprentice doctors. Eventually, in most states apprentice doctors were required to serve three years under an established doctor and then to pass an examination before they were allowed to called themselves doctors. Formal licensing followed slowly in one state after another. In some states it was granted by state boards, and in others by medical schools and medical societies. For whatever reasons, licensing fell by the wayside after 1830, and a degree from a medical school was considered sufficient to allow one to hang out one's shingle. Real state licensing boards had to wait until the 1870s, when they were restored in response to the prodding of the now-more-powerful medical societies. By the turn of the century all the states had some sort of licensing law, although unfortunately, the laws usually included a grandfather clause that allowed many poorly trained doctors to continue in the profession.[6]

As more and more medical schools were established, apprentice doctors slowly disappeared in urban areas. The hospitals that were springing up in the cities and larger towns required more highly trained physicians, giving impetus to formal medical education. The concept of treating patients in central facilities spread slowly, beginning with the first real hospital in Philadelphia in 1752, and gaining ground in the larger cities such as Boston and New York. Initially hospitals were mainly locations for medical treatment for the poor.[7]

Rural America and the frontiers of the westward expansion saw little of all this, either hospitals or trained physicians, until well after the Civil War. People living beyond the influence of the cities and towns relied mainly on old-time medicine, a mixture of home nostrums and the accumulated knowledge of yarb (herb) doctors, midwives, grannies, and medicine men and women.

Thus, before the railroads made their way into the mountains of Appalachia, hardy mountain folk had to rely entirely on themselves in sickness as well as in health. Beyond the town boundaries there were no physicians, no hospitals, no pharmacies. The small family and community graveyards scattered through remote areas of the mountains eloquently testify to the hardships that the very young and the very old faced: there are many tiny headstones, often without even a name. Jack C. Summers of Charleston, West Virginia, wrote: "There is an old abandoned cemetery with beautiful tombstones covered with underbrush and fallen trees on a mountain top overlooking where Sewell used to be. One half the graves in it are children, attesting to the medical attention of the day."[8]

When it came to dealing with sickness or injury, people had few choices. One could rely on the medical folklore passed down through the generations of one's family, some of which usually was derived from the old country, supplemented by Indian medical lore. Or one could turn to someone outside the family who professed medical skills. Many of the latter were older women, usually called *grannies*. Often they practiced general medicine and midwifery. Grannies combined a fulsome knowledge of old folk remedies with an excellent understanding of the value of medicinal plants, and then added a generous helping of plain superstition. It is possible that the superstition served as a placebo, the equivalent of a more modern physician's sugar pill. For example, one might carry a buckeye in one's pocket to ward off rheumatism, or a stinking "fetty" (asafetida) bag around one's neck all winter to ward off colds. Got warts? Apply stump water to the unsightly offender, or rub it with a penny, then hide the penny in a rain barrel. Perhaps the granny knew that left to itself, sooner or later the rheumatism might relent and the wart would probably vanish of its own accord.

Martha Taylor of Meadow Bridge, West Virginia, once lived at Little Sewell Mountain. She wrote: "Most of the babies that were delivered in a radius of ten miles were delivered by my great-grandmother Martha Willis, known by most of her patients as Aunt Mott. She also attended a lot of the sick folks and prescribed teas made of a variety of herbs. She also used poultices."[9]

Midwives are still active in Appalachia; in fact, they have become increasingly popular. Not only are many women opting for home delivery, but also the midwife's bill for her services is very much less than that for a hospital-based delivery.

When it came to knowing the healing properties of plants, the grannies were more often than not on solid ground with their home remedies. Grannies and their male counterparts, the yarb (herb) doctors, scoured the hills to bring back leaves, twigs, and roots to dry or boil or steep—whatever the remedy called for. The resulting powders, teas, and tonics were usually well laced with a bit of moonshine whiskey, and they formed a formidable pharmacopeia. Honey and vinegar were often basic ingredients as well. A granny would mix the two with a little alum to give to a croupy child. Sassafras root, chopped and boiled in water, made a tea that was employed as an annual spring tonic and was guaranteed to chase away the winter blahs. Comfrey tea, concocted from the leaves boiled down in water, was considered excellent for coughs and stomach ulcers, two somewhat unrelated ailments. Plantain and dandelion roots were helpful for those suffering from kidney problems. Elderberries were used to cure almost anything. Wild grapevine root, burned to ashes and mixed with a little water, was touted as an excellent diuretic. Horehound leaves, simmered in water with wild cherry bark, eased a cough then as they do now. Pennyroyal was an almost universal cure, good for everything from insect bites to convulsions. Boneset helped to bring down fevers and mitigated the symptoms of flu, while the willow tree provided aspirin for decades until chemists discovered how to synthesize this most popular of drugs.[10]

Bitters was another cure-all, concocted according to the favorite recipe of the granny or yarb doctor dispensing it. One recipe, for example, called for a mixture of pokeberry, ginseng, wild ratsbane, bloodroot, cherry, sassafras, slippery elm bark, and a goodly portion of corn whiskey. When in later years medicine men brought their shows into the mountains, they grew rich on the sale of bitters, a tonic nicely bottled and colorfully labeled and attested to by happy users. Bitters were always generously laced with alcohol.

Pauline Fisher of Union, West Virginia, who worked as a nurse in Logan, wrote that even during the flu epidemic of 1918 the company doctors in her area battled this scourge with bootleg whiskey:

People died like flies. There was no way the doctors knew how to treat it other than the old ways they dealt with respiratory ailments—mustard plaster to chest to fight by counter irritant and give castor oil. Then they hit on the idea of using liquor. They had outlawed homemade spirits. Logan was forty miles from Williamson [—they would] go there, cross over into Kentucky, into a place called Blue Goose, and get bootleg liquor. It was distilled using corn. It was pure and it was good. The doctors got my father, a minister of the Methodist Church, a great personality, finest-looking, big healthy man; he weighed 280 pounds. He had a big fine overcoat that had a fur collar. He didn't take the flu. . . . Because he was a minister the law wouldn't check him or search him. He could get through the dangerous country into the wild woody hills where the stills were. He could deal with the bootleggers; they trusted him. They aged the spirits, pure corn liquor, in small wooden barrels. They were about one foot high and large enough to hold a gallon or thereabouts. He would buy two barrels and tuck them under this big overcoat, one under each arm, and somehow get them to the doctors. He never did get caught.[11]

Thus for many years, until well after the Civil War, grannies and yarb doctors—along with good corn whiskey—bore the burden of providing medicine in Appalachia. They tended the sick and injured in their areas even as the modern physician tends his practice today, and some of their remedies that strike us as outlandish in fact have survived in some form among the medicines we use. For example, chewing willow bark eased pain because the bark contains an aspirinlike substance that is synthesized in today's pain-killing remedies. Reserpine, which controls high blood pressure for many of us overstressed moderns, is derived from the rauwolfia plant, which Indians have used to induce trances. Ergot, a home remedy and folk-doctor standby for severe headaches, serves the same purpose today after the ergotamine is separated from the fungus. Thus the knowledge of the granny and the yarb doctor has served our present-day drug companies well—not to mention the rest of us.

There were even psychiatrists of a sort in those early days, called hex doctors. I suspect that far back in the hills you might find one today, and I would not be surprised to meet one in a urban area also. The hex doctors played skillfully on the minds of the mountain folk, who were extremely superstitious. Whatever success they had was because the patient was not really sick in the first place and a little homegrown psychology was an effective charm, or because the patient was so mildly ill that time and nature were as much the cure as the faith healer.

The home remedies and those employed by the grannies and yarb doctors were sometimes drastic. The spring rite of sulfur and molasses must have been a little hard to take. Even worse was a headache remedy that consisted of wood ashes steeped in water. (Perhaps it took the patient's attention away from the head and directed it to his stomach.) Turpentine, widely used as an antiseptic and still applied to wounds in many rural areas of Appalachia, must have stung considerably. Perhaps worst of all was a drink that was concocted of sheep manure and administered to cure measles. Even a chest rub of goose grease and onions or turpentine, left on until slightly rancid to cure coughs and colds, could not touch sheep-manure tea.

June Jones of Charleston, West Virginia, who once lived at the north end of Gauley Bridge, wrote about an unsettling treatment that she endured:

> I recalled another medical remedy which happened to me as a small child. First, let me tell you my parents were somewhat enlightened. We children did not have to drink "sheep tea," nor did we have to wear asafetida bags around our necks to ward off germs as many of our friends did. However, I recall having a carbuncle or abscess in my right forearm area. Many salves and hot soaks were used to bring it to a head; nothing worked, not even the slab of bacon which was bandaged to my arm overnight. So I overheard my parents talking in low tones about a "cow plaster," and my father said he would go to the barn and wait for old Spot, our milk cow, to drop a fresh poultice for me. I was terrified. I remember that I crouched, cringed, and cried, but about 11:00 P.M. old Spot obliged, and I was wrapped in the stuff overnight. I vomited and fought to no avail. Later, the bandage was removed and I was bundled off to Dr. Daugharty, the coal company doctor, ... who subjected me to the lance. I carry the scar to this day.[12]

One wonders how anyone survived some of the home and folk remedies. Perhaps a good many did not. Clyde Gray, writing in the *West Virginia Hillbilly*, described one home remedy that went dreadfully wrong. A mountain gentleman named Dave caught a severe chest cold and decided that he needed something powerful to relieve his congestion. He mixed up wood ashes, sulphur, molasses, and lard to make a poultice, then applied it liberally to his chest. It was not long before he felt a fierce burning sensation. He said later that the home-made poultice burned into his skin and that for a good while he picked pieces of raw flesh from his chest and threw them into the fire. A somewhat gruesome tale.[13]

Still, many of these early treatments were founded on sound principles such as the application of heat or cold, as well as the medicinal (and nutritious) properties of the plants employed. I am sure that a bag of hot salt applied to an aching ear did help relieve the pain and inflammation, while cold sliced potatoes wrapped in linen and applied to an aching head surely served the same function as an ice bag. Even the remedies that had little, if any, medicinal value may have worked by the power of suggestion, although some seem a mite on the vindictive side. For example, when bothered by a sty, the patient was advised to stand at a fork in a road or path and chant, "Sty, sty, leave my eye and light on the first one who passes by." Other remedies could be very harsh, such as making croupy children go barefoot all winter.

While neither the granny nor the yarb doctor have com-pletely disappeared from the mountains, their numbers were diminished after the arrival of the smooth-talking medicine men with their road shows. Kings of patent elixirs and skilled in showmanship, they peddled their pretty bottles from a backpack or colorful wagon, mixing home-remedy lore with a good deal of fast talk. A stoic Indian or a pretty girl often helped move the product. The tonics they offered were good for every human ailment under the sun, and were usually heavily laced with whiskey, a ploy that virtually guaranteed good sales and a steady market. These drugs were always accompanied by enthusiastic testimonials from purportedly satisfied customers, such as "I employed ———'s tonic and am today a well man, whereas a few short weeks ago I was pale and puny and close to an end to my life. One bottle and I was on the mend. Three bottles and I could heft one hundred and fifty pounds without effort."

Or a short sad tale would have a happy ending after the touted elixir was administered. A gentleman was taken with pleurisy, with every breath an agony. Six ——— pills were given and warm oil applied to his chest and ribs. Pain was relieved within hours. The second day, six more ——— pills were given, and by evening the gentleman was cured.[14]

Patent medicines were so called because various tonics, pills, and salves were "patented," that is, endorsed, by a king or queen. American entrepreneurs eagerly imported those royally approved drugs and peddled them across the land. Mail-order houses such as Sears Roebuck and Company got into the act; pages of their old catalogues are filled with patent medicines, complete with blurbs and testimonials. From a sales person's point of view, the stuff was pure gold, perhaps simply because of the high alcoholic content. Many a strait-laced gentleman or lady, who would not think of imbibing whiskey, suffered one illness after another that required numerous swigs of bitters. It has been reported that some of the tonics were as much as 47 percent alcohol. Oddly enough, many were more potent than the whiskey of the day.

If they did not contain whiskey, they were apt to contain opium. This, too, did wonders for sales. It also brought about the addiction of a good many people. Until 1920 most opium addicts had developed the habit from patent medicines or medical treatments. One wonders how patients were affected by their tonics, apart from any cure. Laudanum, or tincture of opium, was widely used. It was considered a successful remedy for teething babies, for example. High-strung, nervous ladies were almost instantly calmed by the same elixir. It is rather ironic that our modern pharmaceutical industry was founded, in a manner of speaking, on whiskey and opium.

Grannies and yarb doctors, with their knowledge of the healing properties of plants, and medicine men, with their patent nostrums and sideshows, gave way to the so-called "licensed" doctors. These gentlemen set up practice first on the fringes of Appalachia in small towns. As competition grew, some began to drift into the rural areas. Many simply practiced the medicine that they had learned as an apprentice to an established doctor, who probably was no more formally educated than his apprentices. Many learned their medical skills from experience as they went along, as did the grannies. It is probable that some became comparatively good physicians, limited only by the not-too-abundant medical knowledge of their era. It is also probable that some of them killed about as many of their patients as they cured.

Relatively effective licensing of physicians had to wait until the twentieth century, but even then, poorly trained doctors were allowed to practice under the grandfather clauses. Fortunately, rural folk, especially those in remote areas such as Appalachia, probably did enjoy better health than those who were crowded into the cities and large towns, simply because they escaped many of the big epidemics of cholera and typhoid that swept across America with some regularity. And probably their homemade nostrums were as effective as the treatments doctors in other areas at that time employed, which consisted mainly of the use of mercurials, opium, arsenic and alcohol.[15]

Some of the early treatments lingered long into the twentieth century in the coalfields, as did some of the more potent patent medicines. Pauline Fisher described some treatments in which ignorance was not bliss, however:

> Treating conditions like poison ivy, poison sumac, etc., they used sugar of lead mixed with rubbing alcohol until they discovered that lead was poisoning the system. They also used a baking soda paste with chlorine bleach. . . . This poison if anything was ever terrible. The patients were miserable. . . .
>
> The mustard plasters were a mess. You used dry mustard like cooks used and put it in flour and made a paste with hot water. It was very hard to use. . . . It could blister easily and when used on the chest you had to be very careful to steer clear of that. I learned to use it expertly. I never did blister anyone and I used it years and years.
>
> To sober an alcoholic we used honey. It causes the body to somehow handle alcohol. I wonder why?[16]

The home nostrums also had a long life, and some even are still used in the mountains and elsewhere. Margaret E. Tompkins of Lansing, West Virginia, wrote, "Yellowroot or goldenseal is still used by us for sore throat or mouth. My brother digs the root for us."[17]

Two major medical technologies slowly permeated the coalfields with the arrival of the early doctors: anesthesia, usually in the form of chloroform, and antiseptics, usually carbolic acid, replacing turpentine or kerosene. Even though ether and chloroform arrived on the medical scene in the middle of the nineteenth century, many an early doctor in the coalfields still relied on strong whisky and a stick to bite on, especially during minor surgery. Even for such major surgery as an appendectomy, anesthesia seems not to have made its way to the more remote coal camps until the turn of the century, as witnessed by Lillian Lilly, of Huntington, West Virginia, who lived in a coal camp in the Williamson area:

> People seemed to die young. My very own uncle [was] bent double with severe pain. My grandfather left the tipple in his "bank clothes." You went to the doctor's office for everything, no drug stores or much patent medicine. My uncle Troy was fourteen years old. . . . My grandfather held my uncle. The doctor, with the help of a young lady, operated on my young uncle—with first a big drink of whiskey. I suspect the doctor could have used one, too. Well, the good doctor saved his life, even if it took place on a regular office table.[18]

We can calculate that Mrs. Lilly's uncle, still alive in the mid-1980s, must have undergone his ordeal in the late 1890s, which seems not too long ago for whisky still to have been the anesthesia of choice in the coal camps. Medical advances trickled into Appalachia very slowly; however, the coal-camp doctor need not have felt too backward, since many of his colleagues in less-remote areas still practiced the old-time medicine of purges and narcotics, while bloodletting was slowly passing from the scene. Years after Pasteur proposed his germ theory, physicians apparently still believed that much disease arose spontaneously.

The coal-camp doctor was the first real doctor to arrive equipped with the medical training of the day in Appalachia's remote areas. Initially, these men came from Virginia and from the coalfields of Pennsylvania. They faced a rugged life, and it is a wonder that any of them chose the primitive camps over far more comfortable practices in town.

6

The Coal-Camp Doctor: I

The early coal-camp doctor was educated far differently from his colleagues today. According to Dr. Harry R. Werner, who related his own experiences in the coalfields, a man could obtain a medical degree in 1880 by spending two years in a preparatory academy followed by two years at a medical school. Later the requirements were continually raised, however, first to three years of medical school and then to four. As late as 1902 the only premedical education required was graduation from high school or a teacher's certificate.[1]

Armed with his medical degree, however obtained, the coal-company doctor, like the company store, the school, and the church, was a product of the isolation and the poor transportation in the region where King Coal reigned in the wild mountains and narrow valleys of Appalachia. In its 1923 report, the Children's Bureau of the U.S. Department of Labor stated an additional reason: mining communities were too small to support a regular medical practice; thus, the coal companies had to bring in their own company physicians.[2] The companies devised a system whereby they assessed their miners so much a month, deducting the amount from their pay checks as a checkoff. In a survey the bureau found that single miners paid $0.75 to $1.25 a month while married miners paid $1.50 to $2.00. The fees covered all medical care except for birthing babies and surgical procedures, for which the doctor was generally allowed to charge extra, though usually not much.

Some of the companies gave the doctor the entire checkoff collected, while others held back 10 to 20 percent as reimbursement for collecting the fees. One company in the bureau's report paid its doctor a flat salary of $4,200 a year.

Quite a few of the doctors covered more than one camp, but the camps were almost always no more than several miles apart. None of the doctors in the survey provided care for more than 150 to 200 families. However, since miners' families tended to be large, that could have meant a very busy practice.

One cannot be certain that the Department of Labor survey presented a typical picture, even though it did describe the system. I know that my father, as a post–World War I coal-camp doctor, covered a territory a great deal larger than the three miles that it measured in length. He listed in his records that during some years he traveled as much as 20,000 miles, averaging over 1,000 miles a month.

My father loved to compile statistics. In 1926 he wrote that he delivered 68 babies, while in 1933 he delivered the fewest, only 25. The largest baby he delivered was sixteen pounds, and unfortunately it was stillborn; the smallest was two pounds, and it did not long survive either. He also noted that of the more than 1,000 babies he delivered during his practice in the coalfields he did not receive payment for 245. He also recalled that he delivered 19 babies to unmarried girls, that 6 babies were named after him and two were named after my mother. He also wrote that he charged five dollars for a delivery until 1924, when the rate doubled to ten dollars. By 1931 the charge had gone up to fifteen dollars in some cases and fifty dollars in a few, probably difficult, deliveries, but the fee for most stayed at ten dollars. He also noted in capital letters, "NO MOTHERS LOST SINCE I HAVE BEEN PRACTICING OBSTETRICS." If one considers the conditions in which he often had to work, that was an enviable record.

My father further noted that in 1941 he had 9,196 office calls, 2,316 house calls, 80 deliveries, 15 deaths, 45 accidents, and 327 hospitalizations. During another year he made 426 night calls and averaged 24 office visits every day of the year. There were no relaxing rounds of golf in those days for the practicing physician.

67

Charles M. Mayes, of Deerfield, Florida, related an amusing tale of one company doctor:

> Early twentieth-century coal [mine] operators knew that to be able to operate their mines required a stable, healthy and happy miner, and to lure that man and his family up the hollows and back into desolate mountain mining camps required his having a good family physician as one of the residents.
>
> Most of the doctors hired by the coal companies were compassionate and endeavored to keep the miner and his family healed and happy. But in the mid '20s at Elkridge, West Virginia, a Dr. F—— preferred a good competitive marble game to dutiful family treatment. If, when making his rounds, the good doctor encountered a bunch of men and boys playing "Two Mibs," F—— couldn't resist stopping his rail car and getting into the game, regardless of the seriousness of his patients. Dr. F—— didn't shoot marbles like us boys; he held a large fifty-cent moon aggie between thumb and forefinger, like a Limey pub dart thrower. He was a very good shot but his more-expensive marbles were more challenging for boys to win; they couldn't buy one moon aggie or a bag of glassies; scrip was unavailable during strikes.
>
> Dr. F——, in a competitive marble game behind Black Betsy Coal Company's Elkridge store, was summoned to Rachel B——'s cancerous bedside. Twice promising to come when he finished the game infuriated her son-in-law, who walked down the railroad tracks a half mile with his 32-20 Smith and Wesson tucked in the bib of his many-times washed overalls. Coming upon the game, he asked Dr. F—— if he couldn't come immediately and, getting a negative reply, Dr. F—— found cold steel applied to his right ear, which ended his participation in the game so quickly [that it] caused a mad scrabble for a left-behind choice moon aggie.[3]

The coal-company doctor had to have a strong character, because just attendance at an early-day medical school would try the hardiest of souls, as Harry R. Werner noted in his book. Courses in anatomy and dissection were offered during the first two years, and the bodies used usually were those that no one had claimed from the local morgue. Sometimes poor families of the deceased would agree to let the school have the body in exchange for a shroud and coffin and decent burial. But since the family was unable to pay for a cemetery lot, the body went to the medical college to join others in a cask of preservative fluid to wait until needed. Sometimes cadavers were kept too long in the cask and arrived on the dissecting table somewhat worse for time's wear. Any body part that remained above the preservative would become so dried out and hardened that it had to be soaked in glycerin to soften it enough for dissection.

The bodies, possibly to make for a better fit in the cask, were tied knees to chest, and when brought to the dissecting room, they were simply unloaded on the floor. Apparently somewhat horrified at their first experience in dissection class, the students at Werner's medical school held back, unable to bring themselves to touch these sad human remnants until their professor sternly ordered them to put the bodies on the table and begin dissection.

Twelve students were assigned to each body. Six read from the textbook and six did the actual dissecting. Doctor Werner points out that, since the dissectors had to use their hands to trace blood vessels and nerves through the body and to separate muscles, somebody beside the dissector surely had to turn the pages of the textbook. He also wrote that the smell of the preserved corpse clung not only to his clothes but also to his skin, and no amount of washing could get rid of it entirely. He lost his appetite, particularly for meat, for the first time in his life.[4]

Remembering my own student days in medical school, I am entirely in sympathy with Doctor Werner's reluctance.

When coal-company doctors first appeared on the job, it is probable that miners and their families received better medical care than did the other isolated folk of the region. It is also probable that many, if not most, of these doctors were called on to tend to emergencies outside the coal camps. Until the establishment of regional hospitals, which did not occur until after the turn of the century in West Virginia, the coal-company doctor was the only medical care giver available. Patients he could not treat successfully with his limited means had the alternative of being transported three hundred miles to hospitals in Virginia or Ohio. Since such trips generally were made in a horse-drawn wagon over roads so rough they were a mere excuse, the seriously ill or injured patient was frequently dead on arrival, or even was buried along the way.

Consequently, the coal-camp doctor treated many a serious injury or illness in his office or in the patient's home. Many kitchen tables saw additional duty as a surgical or a birthing table. Often acting as both physician and nurse rolled into one, the doctor would stay with the patient, even if this required several days attending at the bedside.

Lillian Jones of Huntington, West Virginia, who lived on the New River and at Edmond, remembered that, when her second child was born, my father came at night at the family's call and had to spend most of the night waiting for the delivery. The old coal stove petered out during the long hours, and even though it was only September, it was cold in the house. Whenever Mrs. Jones would meet my father after that night, he would tell her that was the closest to freezing he had ever come.[5]

Night or day, in all kinds of weather, the doctor was always on call, often to go to some remote cabin or miner's home perched high on a ridge. On rarer occasions the doctor might treat a trapped and injured miner hundreds of feet underground. Claude Clark of Williamson, West Virginia, recalled that his father, one of the first medical-school graduates to practice in Logan County, West Virginia, used to perform near miracles with nothing other than a few surgical tools, when there was neither a drugstore nor a hospital closer than the large town of Huntington at the time. Doctor Clark "set fractures, amputated limbs, doctored every contagious disease; never had a cripple, never had a cold."[6]

Claude Clark goes on to say that he determined early on that being a doctor was not for him.

My own father had morning and afternoon office hours, but he could be called out at any time, and in some of the worst weather West Virginia's mountains could produce. I never knew him to refuse a house call—except once at midnight in a near blizzard of snow, sleet, and freezing temperatures. A man knocked on our door, asking my father to come tend to his wife. To get to that miner's home, my father would have to drive on a one-lane mountain road where, if the car slipped, he would go straight down the side of the mountain. My father demurred, saying, "Let's wait until morning." The man pulled out a shotgun, saying, "I thought you'd say that!" My father went on the house call after all.

In his memoirs, Doctor Littleton wrote that emergency calls got first priority, but that on occasion the emergency was in the mind of the caller only. He described one sick call when he had to desert his office to drive five miles to a home only to have the mother call the young patient out of a tree he was busily climbing, so that the doctor could see him. The boy did have infected ears, but even so the call certainly was not an emergency. Littleton remarked that often people would brag with a smile that they never needed a doctor, but that if they did, they expected him to come at once. And he observed that then the smile would disappear and a "deadly serious expression" would take its place. He decided that they never considered that he might be out on another emergency when theirs arose, and he was grateful that this situation never occurred in his practice.[7]

Russell Chandler of Paintsville, Kentucky, once worked for the Lake Superior Coal Company in West Virginia. He said that except for serious mine injuries or appendicitis few miners or members of their families ever went to a hospital for medical care. The camp doctor treated just about every contingency either in the patient's home or in his office. Anyone needing his help would leave a request at the mine office or the company store. The doctor would pick up the calls around ten in the morning and again at three in the afternoon. He would keep office hours six days a week, and he always took night house calls. If a patient needed more medical treatment than could be provided in the patient's home, the doctor would hold night office hours and would stay on duty as late as it took to take care of everyone.[8]

Pauline Fisher told an amusing story of a woman who tried to waylay the doctor on his rounds to look at her injured leg. After that doctor picked up his calls at the mine office or store, he would start up the hollows to visit his patients. A person who had a problem, but had not put in a call, would watch for his return down the valley, then approach him with their complaints. The woman, anxious to have her leg examined, saw a man go up to the hollow, and thinking it was the doctor, she watched for his return and accosted him to show him her injury. Pulling her skirt as high as she could, she bared her thigh, turning her leg in one direction or another so that he could get a good look. When she asked him what she should do about her wound, he replied, "Why, consult a doctor. I'm only a sewing machine salesman." Or something of the sort.[9]

Elizabeth M. Weatherford of True, West Virginia, grew up in a coal camp on a mountain near Beckley. She wrote,

One would leave a call at the company store if there was illness in the family, and the doctor would call by the house on his daily rounds. There was always a little black bag from which he dispensed medicines. I never knew what a drugstore was as a child. In those days of typhoid fever and all the diseases of childhood . . . those calls often drug out for days. I am perfectly honest when I say I never recall one being in too much of a hurry to stop and explain what he wanted done.[10]

Doctor Werner described one of his experiences treating typhoid. On a call to a family bedeviled by the illness, he spent the night sleeping with three of the children in the family attic, after attending to two other children with the disease. Only after being assured of their obvious start on the road to recovery did he feel it safe to leave the next day. So many of the children in this family had fallen victims of typhoid that the parents had come to believe they were hexed. Doctor Werner pointed out that their privy was above the spring where they drew their drinking water and that the privy was draining down the slope into that water, bringing all sorts of bacteria, including typhoid germs. He suggested moving the privy to a spot below the spring and filling the old privy pit with lye and earth. He also suggested they boil their water and, in addition, put screens on their windows. With those measures, he assured them, they could avoid typhoid and need not worry about being cursed.[11]

Many a rural and miner family could have profited from that advice.

Thus, the coal-camp doctor had to be a man of many parts, educating his patients in elementary sanitation as part of his practice. In effect, he was a public-health official for the community as well as its doctor. Lillian Lilly recalled: "The whole camp respected and loved 'The Doc,' he being most of the time more of an advisor, problem solver; and somehow he could always separate sickness from a head problem or worry. I know he was not trained for all this. The day for specialization had not arrived."[12]

The ability to improvise was an important characteristic of the company doctor, for he usually was completely on his own, with no nearby hospital or colleague to turn to for consultation. Elizabeth Lambert of Rainelle, West Virginia, remembered how a company doctor saved the life of her cousin, who had been bitten by a poisonous snake:

I recall when a cousin of mine was bitten thirteen times on the feet and ankles; [the doctor] treated him by cutting each puncture wound with his surgeon's knife, bleeding it well, and then, with the help of others, held bottles or vials which contained a small amount of turpentine to draw out the poison. [I suppose the bottles created a suction effect.] . . . The midnight express was flagged and he was sent to the hospital at Montgomery. Although recovery was very slow and painful, it was felt that Doctor Shawver's first aid saved his life.[13]

Mrs. Lambert wrote that the doctor so hated snakes that when someone was bitten he would go the next day to find the snake and kill it.

Doctor Werner described an emergency operation involving a wound that made him ill for the first and only time in his career. After a sleepless night delivering a baby, and still without any breakfast, he was confronted by two men, one of whom held a filthy, bloodstained rag to his face. The other had come along to make sure he got to the "doc." The patient, a lumberjack, was somewhat repulsive. His torn clothing was covered with the kind of sawdust used to cover a barroom floor.

Doctor Werner sat him in his operating chair and removed the rag, to find that the end of the man's nose was attached by only a thin thread of skin. When asked what had happened, the patient's comrade explained that there had been a fight the night before and the other man all but bit off his friend's nose. Before he began to sew the nose back into place, Dr. Werner said he was forced to step to the open window to get fresh air, driven there by the smell of stale liquor and decomposed blood, not to mention sweat. Fortunately, the lumberjack was still so full of whiskey that he allowed his nose to be sewed back together without complaint.[14]

Even as late as World War II, the coal-company doctor's life was far from easy and sometimes a little bizarre. To H. Neil Blount, who lived near Putney as a child, there seemed to be no limits on what a doctor was called to do:

I've heard my older relatives talk of company doctors working two days without going to bed. If it wasn't sewing up a knife wound, he was off delivering a baby somewhere. At times he was even called on to treat one of the mine mules. When Old Doc [was]

... to deliver me, my brother found him in the midst of patching up an old coon dog that had tangled with a bobcat the night before.

Not all company doctors were models of perfection, however. Neil Blount's letter continued:

> When my mother went into labor, she sent my older brother to ride his bike the first five miles to Putney to fetch Old Doc. . . . After a couple of hours, I was delivered by my grandmother. Just as she was finishing cleaning me up, in drove my brother in Doc's old black Ford. The door opened and out rolled Doc with his bag. He was so drunk that he couldn't stand up, so my brother and grandmother half carried and half drug him into the house, where he proceeded to pass out on our sofa till next morning. It was probably the best night's sleep he had had in years.[15]

Makes you wonder what kind of patch job Old Doc did on the coon dog!

In her 1923 report for the U.S. Children's Bureau, Nettie P. McGill was hard on both the coal-company doctors and the system. She indicates how the system, whereby the employer dunned employees for a service, was apt to cause complaints. While some camp doctors were highly regarded, it was charged that others provided inadequate care. Miners and their family members complained that often they had to repeat calls for the doctor several times, and often, when he finally did come, he neglected to follow up on the patient he treated. Worst of all were the claims that the doctor simply sent out pills without determining what sort of illness the patient was suffering.[16]

I do not doubt that there were some bad apples. Perhaps there were quite a few, since the system did allow the doctor to be paid whether or not he treated his patients. Nevertheless, those who have written me about life in the old coal camps—miners, their wives, and their grown children—most often remembered the company doctor with affection and gratitude. Even if the system was flawed and some did not receive proper medical care, one must remember that those company doctors were the first to bring modern medical care of any kind into the mountains of the coalfield. And they did so under conditions that I doubt many modern physicians would tolerate.

Ruth A. Harper of Procious, West Virginia, whose husband

was the company doctor at Swandale during the early 1930s, wrote:

> The salary was $275 a month, on call twenty-four hours every day with office hours from 7:00 to 12:00, 1:00 to 5:00, and 6:00 to 9:00 except Sunday. After one year, one week vacation without pay. The office and drugs were furnished, but no office help. Each employee was charged $2.75 monthly—called [the] check-off plan—for medical care. They could visit [the] office or call [the] doctor as often as they wanted for this fee. [Obstetrical] cases were charged ten dollars and turned in to the company. The Mill was down four to five days a week, so we "caught" many babies. Seven in twenty-four hours was our record, and believe it or not, when the moon changed the babies came. Calls were filled on a motor car [that ran on the railroad track] furnished with driver by the company; sometimes by horseback if the family didn't live along the railroad track. If there were an urgent call, i.e., injury or O.B., the doctor could stop the train . . . he needed the track! The electricity was turned on at 7:00 A.M., off at 10:00 A.M., on again at 5:00, off at 8:00 P.M. In case of serious illness or delivery at night the doctor could request the electricity be turned on. All houses had electricity, so if you got up in the middle of the night, pulled the cord of the bare-bulb light and it came on, you knew Uncle Tom was worse or Lucy was in labor.[17]

I cannot remember if my father, when a company doctor, ever treated mine mules or patched up coon dogs, but he had a few weird adventures of his own. June Jones of Charleston, West Virginia, related the time he went on a call from two old ladies who lived on a steep hillside. To get to his destination, it was necessary to climb a cow path through the yard of another family:

> First I must tell you Thelma's yard was a mess—no grass ever grew in the back because they threw out the dishwater (and the bath water) all the time. Chickens wandered about picking up bread crumbs and other choice morsels. Also hogs came and went at will. Laundry hung frozen from one wash day to another unless needed. . . . Well, you get the general picture—mud and slime four to five inches deep and covered with freshly fallen snow. Through this your unsuspecting Dad went, and down he went like a pontoon. Couldn't even get his footing to rise again. Family members came to help. I understand he used some unique words.[18]

Mrs. Jones has provided me with a clear picture of my poor father, who was a somewhat dignified man, floundering in the muck of that slippery hillside. He faced similar mishaps frequently, as I suppose most company doctors did on their rounds:

> Doctor Frazier was called to Gobbler's Knob, a section of Gatewood, West Virginia. The night was very cold, a blizzard, I believe, and the hour was past midnight. It was a man who was in agony because of a prolapsed thrombosed hemorrhoid. He could not wait until morning. He was moaning with pain. Since there were no streets, let alone house numbers, Doctor Frazier inquired, How on earth could he be found? The man replied that his house was the only one on Gobbler's Knob with a light on and he would turn on every light. . . . And, to please hurry! Poor Dr. Frazier hunted all over the Knob, snow to his knees, nearly frozen, but found no lights, so he finally went home and worried about the poor miserable man the rest of the night. A few days later [the patient] appeared at the office for treatment and when asked why he failed to turn on the lights, replied, "Oh, Doc, after I talked to you I greased my fingers with Vicks and managed to stuff it back inside. I was so relieved I just turned off the lights and went back to bed."[19]

The coal-camp doctor did not enjoy the modern physician's choices in response to a night call, such as directing the patient to the hospital or advising rest, aspirin, and an office visit in the morning. Since the patients had no telephone and therefore no way to discuss their problem or judge its seriousness, the camp doctor had no alternative but to go and see. Even after the miners' hospitals were established in the coalfields, the doctor did not have the luxury of an ambulance or a rescue squad staffed by paramedics. Invariably, the messages relayed stressed urgency, the patient almost always seemed close to expiring, the doctor was needed at once. The doctor had no choice but to go.

On a lighter note, June Jones had another brief comment about my father. She says he carried around a letter sent to him by one of the camp mothers. It was simple and straightforward: "Doctor, my daughter ain't for four months. Oughtn't she?"[20]

My uncle Ralph Frazier also was a coal-company physician for more than a decade, and some of the people who wrote me about their lives in the coal camps remembered him with gratitude. I found the following laconic references to coal-camp practice on different days in his diary:

> 1:30 a.m.: In #36 office waiting for mine accident patient to come out. . . .
> 5:00 p.m.: In #2 mine office to see accident, Negro kicked in mouth by mule. . . .
> 6:00 a.m.: Just now finished a breach on six months premature (colored) upon a hill, pouring rain outside. Still dark outside. House full of flies. Killed flies till ready for delivery. . . .
> Saw 35 patients at home—day 16 hours [long]. Most number of patients in practice so far. . . .

But a day or so later, he wrote: "Saw 38 people at home or house call. Most in 14 1/2 years practice." And on Christmas, this note: "Making calls on Xmas. Made 14 today. Busy all day. 14 calls on Xmas." And on a less weary note: "Delivered 8 1/4 pound male child (white). Child's middle name, named for me (first time). Clayton Ralph Cochran."

7

The Coal-Camp Doctor: II

In the early days of the coal camps, the roads, if they existed at all, were so poor that getting to a patient was almost always a struggle. Travel became even more hazardous and exhausting during the winter months. Thus, the coal camp doctor could face both hardship and danger when making his rounds. Even as late as World War II the more remote camps remained difficult to reach. Martha Taylor, described one mode of transportation frequently employed by doctors, the lucky ones perhaps. One doctor she knew used to make his rounds through the several camps in his territory on a railroad handcar, a vehicle that was not much more than a platform on flanged wheels, with two sets of handles that had to be pumped up and down to set the car in motion. This was fine for the patients who lived within walking distance of the tracks, but not much help when they lived up on the ridges. To reach them, the doctor would have to walk or ride a horse, if one was available.[1]

A single man pumping a handcar along a track must surely have gotten plenty of exercise!

When a doctor riding a handcar met a train, it was up to him to wrestle his vehicle off the track and back on again after the train had passed. He was expected to perform a similar maneuver when he reached his destination. The old pumper

Twenty-ton locomotive

handcars must have been exceedingly difficult to handle. The bicycle type was surely lighter and far easier to move. With four wheels designed to fit on the tracks, it allowed the doctor to pedal his way to his patients. One former coal camp doctor recalled that, even so, driving a handcar could be a risky business, especially if one met a train coming around a curve. Even though coal trains moved slowly, it must have been a hassle to get out of the way. I can imagine the good doctor pedaling serenely along, coattails flying, then meeting a belching steam engine and having to leap off his bike and topple it from the tracks. Perhaps not the most dignified way to answer a house call. And since he then had to walk or ride a horse at the other end of the line to reach his patient, he must have consumed a good deal of time and energy making his calls.

Getting to the patient, however, was only half the battle in the late 1800s and early 1900s. Although medicine was making some extraordinary advances, the coal-camp physician practiced in such remote areas that often he had neither the knowledge of, nor any benefit from, the new findings. Disease in the mining communities ran the gamut of the usual sicknesses of the times. The childhood diseases of scarlet fever, measles, whooping cough, and mumps were ever present. Diphtheria was especially tragic because it thinned the ranks of the very young. Typhoid continued to be a scourge well into the 1930s. And it must have been particularly discouraging to company doctors to have smallpox still claiming an occasional victim after vaccination was an established procedure.[2] Through ignorance or superstition, many mountain folk neglected vaccination, thus ensuring that the disease would keep reappearing. When a smallpox scare arose, many would simply arm their children with a strengthened all-purpose asafetida bag. In early times pesthouses were provided, where smallpox patients had to remain as the disease ran its course and they either survived or died. Those who were fortunate enough to recover were obligated then to nurse those still ill. That task alone did not offer much of an incentive for recovery, since smallpox tends to be an extremely messy disease.

Appendicitis posed a special problem for the coal-camp doctor, for his mountain patients were apt to dub its symptoms "cramp colic," and were inclined to apply home remedies before throwing in the towel and leaving a call for Doc. Unfortunately for the patient, some of the remedies, plus the delay in seeking Doc's help, produced a fatal result. Lillian Lilly's very young uncle was lucky, although, in spite of a hefty gulp of whiskey, he may not have thought so at the time.

Even after the establishment of miners' hospitals in the early 1900s, the coal-camp doctor had often as not to fall back on his own resources. Improvisation was the name of the game.

Dr. Paul Soulsby recalled that, when he replaced an older doctor at Crown Hill Mines the latter left him his scanty collection of surgical instruments with the exception of two, he wished to take into retirement as souvenirs. Both of those were common silverware: a tablespoon and a long kitchen fork bent at right angles. He had employed them years earlier in a lonely mountain cabin to cauterize a woman hemorrhaging badly after childbirth, an improvisation that saved her life.[3]

Mine injuries, frequently severe, presented a real challenge to the camp doctor's skill and resourcefulness. Amputations, ranging from fingers to limbs, were common. One former miner said that the coal miner who has not left a part of himself—a finger, a foot, a hand, an arm, or a leg—in the mine is an exception. Statistics confirm that few miners escaped injury of some kind during their working years. Since 1900 there have been a million such injuries, roughly fifteen thousand each year.

Doctor Soulsby grew up in a coal camp and later went back as a company doctor. His father, a miner, had his right hand mashed between two coal cars. The camp doctor prepared to amputate what was left of the hand, but Mr. Soulsby refused to let him, insisting that without that hand he could not make a living and might as well be dead. The camp doctor relented, treated the hand as best he could, and eventually it became as useful as the left hand, attesting to the skill of the local doctor.[4]

Robert L. Williams, of Charleston, West Virginia, described another gruesome accident, which took place at Stark, where he grew up: "A miner . . . got his hand caught in a coal conveyor belt. He was pulled into the machine until his body became fouled. His arm was pulled apart at the shoulder, letting the end of the bone protrude above the shoulder. Dr. ——— came to the mine about 8:00 p.m., administered morphine, sent the man to Charleston . . . for surgery."[5]

Rose Linsky, a miner's daughter from Charleston, West Virginia, who lived in Putney on Campbell's Creek, added: "The doctor's office was next to the company store. When a miner was injured, they would bring him out in a stretcher to the company doctor for treatment until the 4:00 p.m. passenger train arrived to transport him to the hospital. He was placed in the baggage car along with the mail. Also, bodies were brought in the same way from the morgue."[6]

In the early days an injured miner often had to wait in the mine until quitting time and the "man trip" came to pick up his shift comrades. Between the time of injury and the time of pickup, his companions would have to tend to him as best

Coal cars at South Carbon mine on Cabin Creek in West Virginia

they could.[7] Later, when mining was a bit less primitive, the company doctor could be called into the mine to provide emergency treatment for the severely injured. One company doctor was sent five miles deep into the mine to treat a miner who had come in contact with a live wire. When he returned from this trip on a flat car towed by a mine motor, he was black with coal dust. Dr. Marshall Werner's experience was somewhat different:

> I never cared for people in the mine, and as far as disasters are concerned, they were usually in the form of explosions, and the mine had well-trained teams to bring out the casualties, usually dead.
> As a boy I remember going to school past the rows of bodies on the sidewalk outside the mortuary. Most of the men in that drift died in the explosion, which was enough to blow the large ventilation fans in reverse.[8]

Until the advent of the neighborhood drugstore the company doctor was his own pharmacy. He dispensed what medicines he could contrive, often purchasing drugs by mail order, which were delivered via the coal trains. His little black bag contained samples of his stock, which included sugar pills for the children. Before the turn of the century he would have had little enough in the way of drugs, in any case; his black bag could accommodate his stock with ease. He did have morphine and paregoric, and perhaps even raw opium. He would have had quinine and the dangerous mercurials and arsenic. He had castor oil and aspirin, both of which were employed freely. Then there were the various syrups and tonics, some of which he concocted himself, and which almost always had a strong alcohol base. More thoughtful physicians began to pay attention to sanitation and nutrition, especially the discovery in the early 1900s of the cause and cure of such nutritional diseases as pellagra, an unpleasant condition rarely seen in America today. Pellagra was baffling for the mountain folk because its symptoms would occur in the spring and disappear at the height of summer only to return the following spring with increasing severity. Symptoms could include headaches, dizziness, and burning sensations on the skin, especially of the hands and feet. As the disease progresses and intensifies with each succeeding year, the skin shrivels and turns dark, as though mummifying. The patient may lapse into depression and madness. The cause? Mainly a deficiency of vitamin B complex. Prevention? A proper diet that includes meat, milk, and eggs.

Ruth Harper of Procious, West Virginia, wrote that her company-doctor husband

> requested the store manager to stock oranges; he wanted his babies to have orange juice, crisp bacon, and their formulas—Carnation Milk, water, and Karo Syrup (light Karo, unless constipated, then dark). The mothers thought their new young doctor had strange ideas. One drug he used for colds, especially for babies and children, was Calcidene, a small black pill. He stocked aspirin in many colors and shapes, all five grains. But one patient couldn't get relief with the round blue one, while another needed the square green one—all called pain pills, never aspirin. The summer of 1932, he treated twenty-eight cases of typhoid; all recovered. Almost never an infection after delivery, immune to their own germs perhaps.[9]

Doctor Soulsby told of a colleague who carried not one little black bag, but three. One held his diagnostic equipment, one his obstetrical instruments, and the third his medicines. On one occasion, calling on a woman with stomach complaints, he decided that she was on the hypochondriac side and opted for a mild treatment of bicarbonate of soda. Having none in his supplies, he asked if she had any in the house. She brought him her box of Arm and Hammer, and he wrapped up about a dozen doses in paper napkins. He charged her two dollars for the visit, but later heard from her neighbors that she said he had charged her two dollars for treating her with her own baking soda.[10]

Martha Taylor agreed that the coal-company doctor was his own pharmacy:

> The doctor carried his medications in a little black bag. His medications were in powder form and had a bad taste. There was Calomel, usually six powdered doses, medication measured, I suppose, according to the age of the person. They were in individual papers. They were taken three times daily (no food allowed) and followed by castor oil. When he prescribed cough syrup, he would ask for whiskey (which was a staple in all households); in this he put a few pieces of rock candy and two or three packets of some sort. I think one ingredient must have been cherry because the syrup turned out red, syrupy and cherry flavor (not too bad).[11]

Rose Linsky of Charleston, West Virginia, wrote of her coal-company doctor: "[He] was this easy-going doctor; made house calls, delivered babies, pulled teeth in his office. When kids would see him, me included, we'd say, 'Doctor, will you give me some candy pills?' He would smile, open his bag and give you several pink, sweet tablets, to all the children's delight."[12]

In his memoirs, Doctor Littleton recounted:

> Many patients would have shopping lists of drugs to obtain. I would be given little sheets of paper with several drugs written down and I got the medicine from the drug room. "Pumpkin seeds" were a pill shaped and colored like a pumpkin seed, used for gall bladder and liver disease, and it was quite effective. I had many colors and shapes of basic aspirin, some for headaches, female trouble, backache, etc. It helped to know the families so that I knew which color and shape suited which family for which complaint. I had several liquid cough syrups and other medication for the usual ailments.[13]

There were some, however, who abused Littleton's pharmacy:

A stranger met me in apparent severe pain. He was neatly dressed and stated he was a drug salesman on a company trip and had kidney stones. He walked slowly, with a limp, bending toward his left side and was pushing his back with his left hand. His facial expression was that of torture. He had an X-ray film of a kidney examination and the left kidney was swollen and there was a stone in the left ureter. I wrote him a prescription for a small amount of morphine sulfate tablets, and the prescription could not be refilled. I told him to go to the local hospital. Two days later he was back with the same complaint. This time he had a bottle of bloody urine. He did not go to the hospital as I suggested, since his pain left and he had business in Knoxville. I told him I was not going to help him, but I would refer him to the Lee County Hospital. He next pulled out a wad of money from his back pocket and asked me how much money I wanted to give him a big prescription. Again he was turned down, and I later found out that he was a noted drug addict. He had a bottle of ordinary urine and cut his finger and dribbled blood into the bottle.[14]

The families of miners tended to be large, and that kept the company doctor on the run. In addition, as one doctor recalled, the grannies were almost always sure to turn out for a birthing, not only to "help" but also to offer reams of unwelcome advice. This doctor said that, since he was very young when he started out in the coalfields, the grannies did not seem to think he knew what he was doing. When he arrived on an obstetrical case, he would be greeted by at least three of them, usually the same three, one of whom was a midwife. He said this lady was always "anxious to give me the benefit of her experience" and her favorite trick was to stand close to the bed of a woman in her first labor and mention that she had seen labor this slow, and that the poor woman had died. "Of course, that created quite an uproar!"

This physician went on to describe his method of anesthesia during delivery. He would place gauze in a glass, pour a little chloroform over it and have the woman hold it and breathe the fumes. When she dropped the glass, she had had just enough.

Doctor Littleton presented another view of childbirth:

I recall one house call where the adult miner was having a stone passing down the ureter and he was in agony. He was pushing his abdomen into the table corner to get counter-pain to relieve the prime pain. I prepared and administered the morphine sulfate injection and casually mentioned that this was the only pain a man could have equal to childbirth. Until now his wife had been casual but sympathetic, yet when she heard my statement she burst into laughter, as she had eight of his children and felt he was not sympathetic or helpful.[15]

Mountain terminology was apt to be puzzling to outsiders. One company doctor had difficulty solving the problem of a woman who asked for medicine for her "gravel." After lengthy questioning, she finally admitted, "My water burns me."

He went on to say that while the mountain people were slow to accept a new doctor, once they did they were intensely grateful and "the most loyal patients one will find anywhere," commenting that they maintained their own strong opinions on how to treat various illnesses. He cited a woman who claimed that a company doctor had told her that the best treatment for a child's cold was to place a fried egg on his head, a practice she followed with, she insisted, great success.

It seems strange that a young doctor would choose such a difficult and isolated practice. I suppose for a young physician just starting out the pay was the attraction, plus the ready assortment of captive patients. As we have previously noted, some of the coal companies turned over the entire checkoff to the doctor, while others paid him a flat salary and pocketed the rest. As with the company store, there were times when some companies found the checkoff a hedge against a poor coal market. It is doubtful that coal company doctors ever got rich on their practice.

Most of the letters we received confirmed that miners and their families may have been slow to accept new doctors, but once they did, their respect and admiration knew no limits. Dorlene Benger of Ansted, West Virginia, wrote; "One thing I would like to say is that I think the coal company doctors of forty years ago had a more personal interest in their patients than now; kind of like one big family."[16]

Lillian Lilly fondly remembered a company doctor of her own childhood:

> I can remember how for so long I thought that Doctor Johnson was the greatest man I ever met. I believed with all my heart that he brought all my cousins and all the neighborhood children in his black bag. He always came and, when Cousin Catherine was born, he stayed all night at my grandmother's. He sometimes washed the baby for the first time. . . .
>
> He was always a very kind man. He knew that most of the miners

could hardly read or write. He never expected a lot of pay. But I know now that all the people in the lower camp and also on [the] Hill . . . loved the "Doc" with such a deep respect that he had to feel it. And I know that's why all the coal company doctors I came to know were the best doctors in the world.[17]

Evelyn A. Hopkins of Bluefield, West Virginia, wrote the following memorial to "honor my good friend, Dr. J. Howard Anderson [of the Kingston Pocahontas coal camps], who typifies the coal-company doctor":

> Dr. Anderson was a Pennsylvania native and he started his coal-company career in 1907 when he was about thirty-two years of age. He tried to retire in 1949, but soon began practicing again, and worked at his profession until just before his death on March 16, 1968. He said he had ushered into the world 3,700 "Free Staters." In 1943 he delivered a set of triplets, a rare occasion at this time, and all survived. My mother helped him with the birthing of babies in the homes on many occasions. He delivered the first and the last child in our family of six children. He was courageous and energetic in his practice. He preached preventive medicine, cleanliness, proper nutrition, and good elimination. . . .
>
> When I was three or four years old I was struck down with diphtheria and was in critical condition. At that time we lived at Hampton Roads and I was under the care of another company doctor. When I began to "climb the bed to get my breath," the doctor was sitting there waiting for my final gasp to give me a pill to ease my passing. My dad walked to Marytown to get Dr. Anderson and they walked back through the snow down the railroad tracks, a distance of about five miles. He came in rolling up his sleeves. The other doctor was very angry that Dad had brought him in. Dr Anderson wrapped me tightly in a sheet and forced a tube by mouth into my throat. I remember that it had a string on the end outside my mouth. My color began to return and, after about six weeks with the tube, I recovered. . . . Dr. Anderson told me on several occasions that he had saved my life, but he said it proudly and prayerfully.
>
> When I finished Welch High School in 1935, it was Dr. and Mrs. Anderson who placed money in a scholarship fund that allowed me to go to Berea College in Kentucky. He asked that I not tell it, but I understand that he chose some deserving and needy student to get a start in college each year.[18]

All of this is high praise for physicians who practiced a now vanished brand of medicine in the remote and then wild mountains of Appalachia.

8

Miners' Hospitals

The arrival of the first coal-company doctors was a medical breakthrough for the people of Appalachia; however, for many years those doctors manned a lonely bastion, and both they and their seriously ill or injured patients soon realized more was needed. The nearest hospital was still a long, rough wagon ride or an infrequent train trip away. The end result of a trip to the hospital was too often to arrive dead on arrival.

Just before the turn of the century, a coal-camp physician, who later became governor of West Virginia, went to work to remedy the often deadly void in medical care in the mountains. While his relatives were busy feuding with the McCoys, Henry Hatfield, a child prodigy, had quietly prepared himself for a life beyond the confines of the mountains. Born in 1875 in Logan County, in the heart of the coalfields, Hatfield was fortunate enough to have parents who fiercely believed in education, though getting access to that education was far from easy for the parents and the child.[1]

It is said that Hatfield went to school riding behind his mother on a white mare. He would stay with an aunt during the week and return home every Friday, when his one pair of overalls was washed. One of his relatives said he was "smart from the boots up," and indeed, Hatfield graduated from college in Ohio before the age of fifteen. He went on to study medicine at the University of Louisville, Kentucky, becoming a licensed physician in the state of West Virginia at eighteen. His thirst for medical knowledge seems to have been far from quenched, however, because after practicing as a railroad and coal-camp physician for several years, he acquired a second medical degree from New York University. He was still only in his early twenties.[2]

Hatfield's stint in the coalfields as a company doctor motivated his future efforts to establish miners' hospitals. Along with his coal-company colleagues, he had had to perform major surgery on kitchen tables and deliver babies in dirty beds. When hospitalization was utterly necessary, he had to send patients to the nearest hospital—200 miles away. And like his colleagues, Hatfield found that all too often his patients did not survive the trip. Neither the railroad companies nor the coal-mine operators whom he approached for help were interested in establishing regional hospitals to care for the men, often severely injured, who worked in their enterprises. Even when Hatfield compiled statistics proving that it was in their camps that the most serious illnesses and injuries occurred, he met with complete disinterest.

However, Henry Hatfield was not easily discouraged. Dedicated and stubborn, he switched his efforts to the state legislature, where he finally prevailed. In 1899 the West Virginia legislature passed an act establishing three public hospitals in the coalfields, to be called "the Miners' Hospitals." That designation did not bar others from being admitted, but they had to pay, whereas those injured as a result of their occupation as a lumberman, railroad worker, or miner were treated free of charge. The hospitals were restricted to acute care; infectious-disease patients were restricted from admittance.[3]

Miners' Hospital No. 1 was built at Welch at the junction of Brown's Creek and the Tug River in a floodplain. As a result, some of the hospital grounds were little better than a swamp. The day of the hospital's opening, Brown's Creek was in flood, and the water was so high that the two hospital nurses were unable to get to work.[4]

The hospital's location proved to be awkward for yet another reason: it had been built some distance from the railroad depot. Patients made the trip from the station by wagon on a rutted track, then by stretcher on a footbridge over Brown's Creek. Fortunately, the railroad later provided a special stop

Governor Henry Hatfield of West Virginia

Miners' Hospital No. 1, Welch, West Virginia

for the hospital. However, even if the hospital's location left something to be desired, it can be imagined that Doctor Hatfield, as a member of the board of supervisors, felt his efforts had been rewarded.

Miners' Hospital No. 1 was not large, and indeed none of the three hospitals initially had more than thirty or forty beds available. During the first year of its operation No. 1 treated 280 patients, although the two-and-a-half-story building was designed to care for only 25, and 40 in an emergency. There were times, however, when as many as 65 patients were accommodated. The annual report of the state control board recorded that 38 of the 280 patients in the first year were railroad men, 146 were miners, and the rest were what was then called "paying patients" (those not hospitalized as a result of an occupational accident). The report listed 147 patients as discharged and 27 as having died, and the rest, we can suppose, were still hospitalized at the time of the report's publication.[5]

Initially, the staff of Hospital No. 1 consisted of a doctor in charge, two nurses, two orderlies, an ambulance driver, a night foreman, and an electrician. A matron also was listed in the first biennial report. She probably took care of housekeeping details. Even though in its early days conditions at the

Welch, West Virginia, train station

hospital were somewhat primitive, it was within comparatively easy reach of the coal-camp doctors and their patients
in the region that it served, and for many years it served them
well.[6]

Doctor Hatfield served as chief physician and surgeon in
No. 1 for thirteen years. He recorded that he started out from
his home at 5:00 A.M., took the train into Welch, then walked
the remaining one and one-half miles to the hospital. At 4:00
in the afternoon he would leave the hospital and take the train
to the coal camps, where he still served as a company doctor,
and there he made his rounds on horseback, often not getting
back to his home until around midnight. He said with obvious
pride that except for a few breaks he kept up this schedule for
all thirteen years.[7]

Miners' Hospital No. 2 was constructed about 1910 on a
lovely site above the New River near the tiny town of
McKendree, which had a population of less than fifty souls.
The hospital was on several acres in a stately grove of trees on
a plateau along the mountainside. Whoever chose the location

must have had a keen eye for beauty. In its 1910 report, the West Virginia State Board of Control stated that the hospital, situated above the railroad, had a lovely view of the New River valley below. It praised the cool, fresh air of the altitude and the peace of the surrounding forests, saying that all this should greatly benefit the sick and injured.[8]

The town of McKendree consisted of little more than a small railroad depot, a tiny store, a boardinghouse, and several homes. It was so small a town that its post office was located in the hospital, where no doubt most of the mail went anyway.[9]

Miners' Hospital No. 3 was located in the hamlet of Fairmont, where the local residents had raised the money to buy land for the hospital. Like the McKendree hospital, No. 3 possessed a wonderful river view overlooking the Monongahela. During its first year it took in almost two hundred patients. The first superintendent stated in his annual report that it was the only hospital in the area "from Cumberland [Maryland] to Wheeling, from Clarksburg to Connellsville [Pennsylvania]."[10]

As coal towns began to grow, other non-state-supported hospitals were constructed such as the one at Beckley, a town close enough to the mines to accommodate coal-company patients. In this private endeavor, coal operators provided for the miners' care by deducting a monthly assessment from their paychecks and turning it over to the hospital. Thus the miners and their families had prepaid hospitalization for all noninfectious illnesses.

In his column in the *West Virginia Hillbilly*, "Remembering the Past 100 Years, and Then Some," Clyde Gray described a hospital visit that he made as a boy in 1914 to have his tonsils out. He said the Sheltering Arms Hospital was a large brick building sprawling across a hillside above the river. Since it was summer, patients were sleeping on the large porch, and he was assigned one corner to await his operation. Nearby was a young man who had been shot in the abdomen. Apparently, the damage was so severe that the doctors did not operate. Gray said that the young man "lay there screaming and crying for food that he could not have. His body continued to waste away and infection soon took his life." When later in life he passed where the hospital had once stood, Gray could still hear in his mind the man's tormented cries.[11]

Like so many other miners' hospitals, Sheltering Arms was

closed in the 1920s and torn down soon after. Miners' Hospitals 1 and 3 have been replaced by newer structures. No. 2, now in ruins on its site above the New River, served as a home for elderly blacks after it had ceased operation as a hospital.

And so another chapter in the history of coalfield medicine came to an end. But like the coal-company doctors, miners' hospitals are remembered with a great deal of affection and personal gratitude.

9

The Nurses

Nursing, as a profession, dates back to the example provided by Florence Nightingale during the Crimean War of 1854.[1] Nursing, as a skill, has a history as old as humankind, rising from the need to care for children, the sick, and the injured. Intelligent and energetic women developed a body of medical knowledge through practical experience caring for their own families and were often called on to help others outside the family circle.

Religious orders provided nursing care for the poor, and by the eighteenth century many Catholic and Lutheran nuns staffed the early hospitals as nurses. As the number of hospitals increased both in America and Europe, the nursing orders could not meet the demand and secular nurses were recruited. Still, practical experience was the only training for all but the obstetrical nurses, who were given some instruction as they accompanied doctors into the home as the first home-care nurses. It is interesting to note that nurses, both male and female, served aboard navy ships along with women cooks and laundry personnel.

The status of nursing as a profession arose with women's long struggle for equality. Florence Nightingale was an early and confirmed feminist who believed that women should have the right to work in any occupation open to men. She also believed that it was high time women had access to political power. When she was appointed superintendent of the Female Nursing Establishment in the British hospitals in Turkey during the Crimean War, she proved extremely adept in establishing and running hospitals. And the efforts of her nurses to save the lives of wounded soldiers were so successful that the death rate fell dramatically. Thus the importance of nurses was finally, if grudgingly, recognized by the medical establishment.

"The Lady With the Lamp" did not stop there. Nightingale went on to battle the apathy and stupidity that kept the British Army Medical Services stuck in the Dark Ages. A fierce believer in sanitation, she wrote reports detailing the measures necessary to prevent such diseases as typhoid and typhus and the importance of attention to the general health of the troops. She revolutionized the management of hospitals in and out of the military. And for the first time, she defined nursing in her book, *Notes on Nursing: What It Is and What It Is Not*. From there she went on to establish the Nightingale Training School.[2]

Florence Nightingale's efforts were not lost on the women of America. The Civil War wounded were helped by Clara Barton, who went on to organize the American Red Cross; Mother Bicerdyke, who would search the battlefields at night to care for wounded soldiers left behind; and Dorothy Dix, who organized a corps of Union Army nurses.[3] By 1870 there were American nurse-training schools modeled on Nightingale's example. Probably the first formal school and Hospital training program for nurses was instituted at Boston's New England Hospital for Women and Children under the guidance of Dr. Marie Zakrzewska between 1862 and 1902. Other hospitals followed her example, and by the turn of the century there were over 400 nursing schools in this country.[4]

Meanwhile, in the isolation of the coalfields the grannies and the midwives held their ground for many years. The establishment of the miners' hospitals early in the twentieth century diminished their role, but in some areas they still did not disappeared entirely. Doctors still had only these untrained, though fully experienced, women to aid them in caring for the critically ill or injured, to free them from the bedside vigils that often kept the doctor tied to one patient for days.

A few years after the construction of the McKendree Miners'

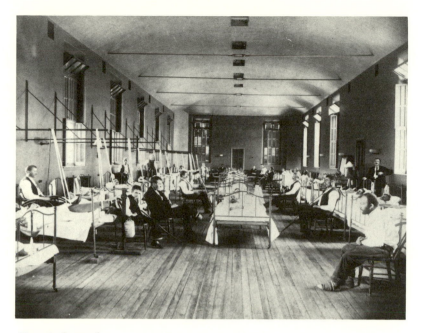

Hospital ward

Hospital in 1910, a second building was added to house the McKendree Hospital Training School for Nurses. The State Control Board had set up standards for student nurses and their training program, insisting that all who were accepted be of high moral character, intelligent, and in good health. The course lasted two years and included practical work as well as study. The trainees, when not in class, worked among the hospital patients. They were paid ten dollars a month, with which they were expected to pay for their uniforms, textbooks, and the like. Those on the day shift worked and studied from 8:00 in the morning until 8:00 at night, while the night shift took the remaining twelve hours. One gets the impression that there was more time spent on the practicum than on study. In the course of the day, the trainees were given one or two hours of free time and could go off duty from two o'clock on Sunday afternoon until eight. They were granted an annual two-week vacation.[5]

I have a very special interest in Miners' Hospital No. 2,

which was dubbed McKendree No. 2 locally, and is now a vine-covered ruin above the New River. My mother, Nina Francis Toney Frazier, graduated from the Training School for Nurses in 1915, five years after the school was established. She used to shake her head in disbelief when recalling that the students worked that long twelve-hour day and then attended classes at night. She also remembered that all of the hospital beds were always occupied and that many of the patients had suffered severe injuries. Limbs had to be amputated, bones set, wounds sewed. The wild-and-wooly downriver town of Thurmond kept the hospital supplied with gunshot- and knife-wound victims. Though not unique in the coalfields, Thurmond had a reputation that seems to have rivaled that of rambunctious towns in the wild West.

Chief among the serious illnesses that brought patients to McKendree and its sister hospitals was typhoid because of the poor sanitation in the crowded coal camps and in the mines themselves. Unfortunately, the disease's victims lived or died more by the whim of chance than because of any treatment that the hospitals could provide. In the days before antibiotics there was little medicine could do.

But all that could be done, the nurses did. They administered aspirin for the discomfort and the fever, forced plenty of liquids to prevent dehydration, and also gave alcohol rubs to bring down fever. They would feed the sick, often spending a good deal of time and patience on one patient in an effort to nourish and preserve strength. To tend those patients, they would don special gowns and wear rubber gloves and gauze face masks. Those who did survive could thank not only Lady Luck but also the dedicated nurses whose care often made the difference.

Typhoid was not the only scourge of the region. Tuberculosis was rampant in the coalfields, possibly due in part to the smoking slag heaps, described by Pauline Fisher:

> Logan was only 475 feet above sea level at the courthouse—
> Down in the hills. Damp, and when the slate stacks would catch
> fire and smolder days, months, years, the process turned slate
> [slag] into a product they called "red dog." . . . When the smoke
> got hovering in a pall over the slate dumps, as they called them,
> you couldn't see the sun for as much as six weeks at a time. So lots
> of TB.

The Nurses

You could go to Pinecrest, the TB hospital, stay all day and spend only fifteen minutes in each room, and not visit everyone there from Logan in one day. The treatment for TB was really something. At night, the patients told me, they would rub their bodies from head to foot with musterole [mustard oil]. Their muscles would ache so. With all nursing personnel rubbing that much musterole, well, maybe they used gloves. I dread to think of their hands.[6]

Pauline Fisher went on to say that castor oil, evidently in quantity, was part of the treatment of tuberculosis. She adds some of her other experiences as a nurse:

I would be doing a special-duty case. . . . You could hear ambulances rolling in with mining emergency cases and regardless of who you were working for, doctors would step out into the hall and call out, "A little help here!" They didn't even say "please." No time for that. And all available nurses would run to help. This was normal procedure for any day. . . .

I remember one patient crushed completely in a slate fall. He was in traction and bandages, with legs up in the air, straight up, arms straight up—funny looking setup. Only his mouth and nose was not bandaged. He didn't make it, but they tried everything to help him. He was simply crushed. These were pitiful sights and wonderful workers to lose. . . .

I delivered more babies at home for ten years than all three [miners'] hospitals put together. Successfully, and never lost a patient. I would go on location, set up for delivery and manage the labor down until the pains were five minutes apart and call Doctor for help. I never had him more than five minutes to thirty minutes before birth in all cases for the whole ten years.[7]

She adds amusing anecdotes about some of the physicians:

Once he [the doctor] was called up one of the hollows to deliver a fourteen-year-old girl who didn't even know the name of the baby's father. They were a very illiterate set of people. Doctor got vexed because the giggling girl couldn't cooperate.

He said, "I can't deliver this baby unless you get in better shape." She said, "I don't know what kind of shape you want me in." He said to get in the shape you were in when you got this way. She said, "You mean I have to get in an old Ford car and stick my feet out the door?"

Another similar incident just as bizarre as that one. He held up the baby and said, "You have a fine boy."

She said, "Just one? We got into something twice." She didn't
know the baby's father's name, either. So goes life in the hollows![8]

Ms. Fisher told the treatment for patients with venereal
disease, which was rather drastic:

> There was so much venereal disease in Logan County the Federal
> Government put up a disease-control clinic, their office in the
> courthouse in Logan. If patients would not go for treatment, they
> put them in jail. The men had to give their contact. They put her
> in jail also and would handcuff them together and march them
> down main street to get their shots—606 they called the shots. One
> weekend one "weed monkey," as they called the girls, infected
> thirty-five men. You can bet a man will go to a doctor on Monday—
> the pain is so intense it's impossible to stand.[9]

The drug called 606 was arsphenamine, a yellow powder
that was almost one-third arsenic. Introduced around 1910 to
treat syphilis, it replaced mercurial drugs for the most part
because it was exceedingly effective. Arsphenamine was replaced in turn by bismuth compounds and a new drug,
mapharsen. When penicillin was introduced at the time of
World War II, it was a giant step forward in the treatment of
venereal diseases, far superior even than the sulfonamides
then employed to treat gonorrhea. When syphilis affected the
nervous system, doctors applied fever therapy, whereby a fever as high as 105 was induced and maintained for several
hours. Happily, medical progress was matched by social advances, and infected couples were no longer handcuffed together and marched down Main Street.

As Pauline Fisher noted, nurses in the coal regions saw a
good deal of the seamier side of life. And nurses then, as now,
were greatly undervalued. They contributed more to the recovery of their patients than most physicians and the patients
themselves realized. Conversely, the training that nurses received was often of great value in other avocations, such as
being the wife of a coal-camp doctor; as my mother related:

> There were many interesting experiences during our years in the
> coalfields where my husband was the only doctor. I helped deliver
> babies. I took fish bones out of tonsils. I even delivered a baby once
> when my husband was out on another call. I helped my husband
> on night calls over cliffs and in shrubbery where there were bears

and snakes around. I recall a miner threatening my husband's life. Once my husband saw one man shoot another man and then dare my husband to go help the wounded man. He saw a young boy stab his father several times. He was threatened because a child died. It was a tough life then; the depression years, three children in school, no money. We saved a little and lost what we had. No one had any money to buy food, much less money to pay the doctor. My husband bought his own drugs to treat people. He walked for the first six years of practice.[10]

There is no doubt that my mother's training as a nurse stood my father in good stead. I am also sure that the tough regimen she went through during her training at McKendree taught her fortitude, responsibility, and patience, for she went on to say: "My husband never had more than one or two vacations in all that time. As many as five years would go by without his seeing his parents. If he ever left home to go fourteen miles away to the drugstore, he would return to find a dozen patients on our porch demanding to know where he had been and insisting that he was needed immediately." . . .[11] I too can clearly remember that porch full of impatient patients!

Paul J. Nyden interviewed Mabel Gwinn, who had entered McKendree Training School for Nurses in 1921 at the age of seventeen. She described all the things that trainees were expected to do over and above learning nursing care. Although they had an orderly to scrub the floors, the young nurses-to-be often had to clean the bathroom and beds and keep windowsills and woodwork spotless. They had to soak bedpans in the bathtub for hours, which then meant scrubbing out the bathtub. The doctor in charge, Mabel Gwinn explained, "was very strict about the cleanliness of the hospital."[12]

To accomplish all this housework plus care of the patients, the trainees worked a day shift of ten hours and a night shift of twelve. If they did anything wrong, or what the supervisor considered wrong, they would lose those two precious off-duty hours and have to work all twelve.

Nurse Gwinn recalled that the patients came by train day and night, and that often the hospital staff would clean them up, treat them, and then turn them loose. More serious cases, of course, would be kept, perhaps as many as twenty or twenty-five a week.

Mrs. Eva Ruth McKean of Marlinton, West Virginia, wrote

of her experiences as a coal-camp nurse at a Consolidated Coal Company mine. Shortly after she graduated from nurse training, she went off to visit her brother who lived in the coalfields of McDowell County, and there she got a job as a coal-company nurse. There were two young doctors with the company, and she would assist them in special clinics, although not during their office hours. Among her jobs was to keep surgical supplies in stock and give health lectures every month in each of the classes of the camp's school. She also taught a first-aid team of older girls and inspected all the schoolchildren for health problems. Those in whom she found an illness were sent to the doctors.

Nurse McKean's company seems to have been far advanced in the medical care it offered in the early 1930s. The doctors conducted three clinics: a prenatal clinic, a well-baby clinic, and an immunization clinic. For the prenatal clinic, McKean would obtain a list of all the pregnant women, visit them once a month to distribute literature, urge them to attend the clinic on a regular basis, and check to make sure that they had everything necessary for delivery at home when the time came. Only women who had possible complications were sent to the hospital.

At the well-baby clinic, babies were checked when a month old and every month thereafter until the doctors could be assured that they had made a good start in life. Mrs. McKean said that the mothers were immensely proud of their babies and would bring them to the clinic "all shined up."

The immunization clinic was initiated after a severe typhoid epidemic several years earlier. The clinic was held at regular intervals to be sure that children and newcomers were immunized against not only typhoid but also smallpox and diphtheria. Most of the camp inhabitants cooperated and there were no more cases of typhoid as a result.

One of her duties, Mrs. McKean recalled, was to join the mine superintendent and one of the doctors on regular tours of inspection around the camp. They "inspected the yard and the surrounding areas of each home, approximately five hundred homes. It took the entire day, walking both winter and summer. If any areas were found unsanitary, the house number was given to the nurse and she checked back and advised the family. Most families were very cooperative."[13]

Mrs. McKean summarized her McDowell County experience

Union relief day, Matewan, West Virginia

favorably: "There were very few injuries at the mine and [they] were seen only by the doctor, who sent them to the hospital, if necessary. All employees were required to take first-aid training, so accidents were at a minimum. Since prevention of disease was highly stressed, I did very little active nursing while with this company."[14]

Apparently when the Great Depression hit and the mines were closed for all but one or two days a week, the nursing service was suspended. But Mrs. McKean's nursing skills were needed again in another camp:

> About a year after leaving the company, a doctor from a small coal company came to me and stated that he had thirty cases of typhoid fever in his camp and asked me if I could help him for thirty days. He furnished the car and the gasoline.
> I went into the community early and stayed late—all the patients were in bed and very sick. . . . I trained one member of the family to take temperatures, give sponge baths for high fever, and follow the diet list. Also keep a daily record. We began with a liquid diet consisting largely of buttermilk and juices each hour while awake, and water between. As they improved, soft foods were added.

Immunization clinics were held at designated places where typhoid injections were given—all were completed and the people were very cooperative. At the end of thirty days all patients had recovered. One eighteen-year-old boy had a serious heart condition before he contracted typhoid fever. He recovered but later died of heart complications.

The doctor was even pleased and I must say this was one of the most interesting and rewarding pieces of work of my nursing career.[15]

As the mines slowed production during the Great Depression, and miners either lost their job or worked only one or two days a week, hunger cast its shadow over Appalachia. Men had a hard time getting enough food for their families; nursing mothers could hardly feed their babies. Pauline Fisher described an incident of hunger:

I'd nursed a neighbor across the street from us with her new baby. People nursed their babies natural then. . . . The baby began to cry all night. I was sleeping on that side of the house, and it was such a pitiful cry. I felt it was hungry. In desperation one night I called her and said, "Would you like for me to bring the baby over and rock it to sleep and let you get a bit of rest?"

She said, "Pauline, it will probably spoil it rotten, but I'm so tired I don't care if you do."

I'd do that and it wasn't hard to realize it was hungry. I politely went to town, got me a little bottle and a nipple and a box of Cream of Wheat. We had a cow, so I'd cook Cream of Wheat real thin in the milk, sugar it. I cut a hole in the nipple and I'd feed that baby the whole eight ounces and it would go sound asleep.[16]

Like the midwives and the grannies, nurses in the mountains had to use their heads as well as their hearts.

10

The Coal Miner's Nightmare: Black Lung

Hawk's Nest Tunnel: It was more than a tragedy, it was an incredible disgrace for the state of West Virginia. However, it set in motion efforts to provide tougher safety laws and stricter enforcement measures to protect miners against the hazards of coal dust, and other workers from the detrimental effects of various occupational dusts.

Hawk's Nest Tunnel was an ambitious project begun in 1930. The plan was to build a tunnel, almost four miles in length, through a mountain at Gauley Bridge to divert the waters of the New River for a hydroelectric project. The work was subcontracted out to a Virginia construction outfit.

Two thousand workers were hired, the majority of them black. Shanties were built to house them in two segregated and dilapidated camps, and the work began. It was discovered early in the project that much of the rock to be blasted out contained silica. As they got closer to the center of the mountain, it was found in an almost pure form. No one paid much attention to this fact, and the workers continued doggedly to blast and shovel their way through the mountain.

Within a month of the beginning of the project, workers involved in the blasting, digging, and removal of the rock began to complain of shortness of breath. Some were sent to the hospital with a diagnosis of pneumonia, and some died, but since pneumonia was a common health problem, especially among blacks, no real attention was paid to the growing number of ill workers—only to the inconvenience of finding replacements. No attention was paid, that is, until some white workers also became ill.

In spite of that, work went on. By now so many workers were dying that the company had to take out a burial contract with an undertaker. In the end at least 476 men died from a disease referred to as "tunnelitis." Many of them lie in unmarked mass graves near the New River. There was no compensation for their families; no recognition that they had died as a result of ignorance at best, or criminal neglect and greed, at worst. All of the men sickened and died within the eighteen months it took to complete the tunnel.[1]

It was the very shortness of the time span, plus the number of workers who succumbed, that finally drew national attention. In 1936 the U.S. Congress appointed a committee to investigate. Surviving workers testified to conditions in the tunnel. One man stated that the dust was so thick he could see only several feet in front of him. Another said it was so thick that he once bumped into a steam shovel. As the workers dug deeper into the mountain, ventilation became totally inadequate. The testimony of one witness illustrated clearly why the dust was so thick and why the men were continually exposed. He stated that, when he went to Hawk's Nest Tunnel, he worked as a driller on the bottom bench to drill dry holes one to twenty feet straight down, and that the dust this created was especially heavy. He said that they did not use water during the drilling because dry drilling was about three times faster. He also testified that they were continually urged by the foreman to hurry and not worry about falling rocks during blasting, just keep on working.[2]

The dry drilling created far more dust than would wet drilling. Also, in the rush to complete the tunnel, the men were forced back in to work before the waiting period required by law as a safety measure after blasting. This practice exposed the workers to vast amounts of dust that had not had time to settle. No face masks or respirators were provided. In the end, not even the congressional investigating committee really knew how many men died at Hawk's Nest Tunnel, for many who became sick were summarily discharged without medical care and may have returned home to die.

The Coal Miner's Nightmare

Alas, those who ignore history are doomed to repeat it. As early as the 1700s it was known that breathing dust containing silica could cause lung disease. Several centuries ago, knife sharpeners, pottery workers, and people who ground fine points on needles were frequent victims of what we now know was *silicosis*. It is the fine particles of silica dust such occupations produce that does the damage.

Like silicosis, black-lung disease has been around ever since men tunneled deep into the earth after coal. Physicians have been aware of its existence for many years. The condition of progressive breathlessness and racking cough was called miner's asthma in the early nineteenth century. The cause then was not known. Nor in this country was much effort directed at finding out why coal miners developed and died from this condition.[3]

In that respect coal-company physicians were tragically remiss, if not derelict to their patients, caught as they were between the coal companies, who wished to have black lung ignored, and the miners, who suffered the consequences. Many physicians were content to condemn the silica sometimes found in coal as the culprit, thus labeling the disease as silicosis, while coal-mine operators assured everyone that new methods of ventilation would solve the problem.

Even when coal dust was finally recognized as the cause, the condition was medically denied for several decades, effectively cutting crippled miners off from any compensation. Economic powers in the United States combined with political powers to keep mine operators and state legislators blind to the plight of stricken miners and to the deprivation suffered by their families when black lung destroyed their ability to work or killed them. As late as the 1960s coal companies and state lawmakers refused to admit that coal dust had anything to do with the respiratory problems of miners. They even accused miners of bringing on the emphysemalike disease by heavy smoking. Smoking, of course, does aggravate and hasten the deterioration of coal dust–affected lungs, but it does not cause black lung.

In contrast, in the 1940s the British not only recognized black lung as an occupational disease but also acknowledged that coal dust was the culprit, and they provided compensation for disabled miners, following extensive research into the changes coal dust causes in their lungs. The British also led the way in developing preventive measures. Neither American medical science nor the American coal industry paid any attention to British findings.[4]

It took an explosion, the Farmington, West Virginia, disaster in 1968, to galvanize Congress to recognize the plight of miners in general—both the safety issues and the threats to their health while on the job. In 1969 Congress, after hearings indicated that over three hundred thousand miners suffered from black lung and thousands had already died prematurely, moved to bring black lung within the sphere of compensable occupational diseases. The Federal Coal Mine Health and Safety Act of that year produced regulations for coal-dust concentrations in the mines. Unfortunately, as for other well-meant legislation, enforcement has left a good deal to be desired. Even today, cuts in the United States Mine Bureau's budget may leave enforcement spotty at best. There is an uneasy feeling among today's miners that they too will complete twenty or thirty years in the mines and end up scarcely able to draw breath, even as did their fathers and grandfathers.

In the early 1950s, even before the Farmington tragedy, the United Mine Workers' Welfare and Retirement Fund began to offer both treatment and rehabilitation. Later it joined the newly formed West Virginia Black Lung Association to begin a campaign, eventually successfully, in the West Virginia legislature to bring black lung under Workers' Compensation. It took a walkout from the mines and a march on the state capitol to ensure passage of the legislation.

Several physicians in West Virginia, especially Donald Rasmussen, I. E. Buff, and Lorin P. Kerr, deserve immense credit for their research and work on black lung. In spite of criticism from many of their medical colleagues, and despite the indifference and even hostility of mine operators, they initiated a testing program and conducted a study of coal miners with the disease. The results presented the first real measure of how severe the problem was in West Virginia, sparking the development of the militant Black Lung Association.[5] By that time tens of thousands of miners had been disabled, forced to live out what was left of their lives in pain and poverty, and thousands of widows and children had been reduced to despair.

With the problem finally addressed in the coalfields, the

The Coal Miner's Nightmare

U.S. Congress became aware that so many miners suffered from this debilitating disease that the coal companies and the coal states could not provide even partial compensation. It is estimated that at least 365,000 miners were affected, but the numbers remain uncertain because the death certificates are misleading.[6]

What is black lung? Why is it peculiar to coal workers, specifically those who work with bituminous, or soft, coal? Previously called *anthracosis*, or simply miners' consumption, black lung is now tagged as *coal workers' pneumoconiosis* within the medical community. It is one of a group of similar lung conditions, caused by different dusts, that all come under the broad term penumoconiosis. Black lung produces a condition similar to silicosis, asbestosis, byssinosis (brown lung, caused by cotton dust), and farmer's lung (caused by moldy hay). For example, rock-quarry workers and those who work in silica-crushing operations develop a disease so akin to that of the coal miners that for a long time physicians believed that miners suffered from silicosis. However, black lung is caused by the coal dust itself, with or without the addition of silica dust. We now know that *miners' asthma* is quite a different condition from black lung. It is a result of allergy to some substance in the coal dust or to fungi found in damp mines.

By whatever label, black lung is a tragic disease, made even more so by its preventability. It takes two forms, simple and complex. In its simple form, called Stage One by the miners, dust particles adhere to the bronchi and lung tissue. At this phase of the disease a miner should be removed from further exposure to coal dust while he can still function normally. Respiratory function may be mildly affected, and further exposure could bring on Stage Two or Stage Three, which miners indicate is the stage at which black lung develops into its complex form. To judge whether a miner should receive compensation, the government characterized Stage Two as involving small black masses in lung tissue. At Stage Three such masses have increased in size, sometimes becoming as large as a fifty-cent piece. These stages of miner's pneumoconiosis are marked by actual changes in lung tissue. The masses of black tissue formed in Stage Two may or may not, physicians believe, progress to the larger and more devastating masses of Stage Three. But once Stage Three is reached, the disease is progressive; the black masses may increase in size until pressure is exerted on the heart and pulmonary vessels. The end result is disability and death, the latter frequently the result of heart involvement.[7]

The initial symptom of black lung is coughing, which will increase in violence with continuing exposure to coal dust. The sick miner becomes shorter of breath as the disease progresses. Eventually he becomes weaker and must stop every few feet to rest, gasping for air. He may suffer from chest pain, and his lips and ears may take on an unhealthy blue color as the normal flow of blood oxygen is reduced by the decreased capacity of his lungs. Chronic bronchitis, emphysema, pneumonia, and tuberculosis are frequent complications accompanying the destruction of lung tissue. Starved for oxygen, muscles grow weak, and any exertion brings on a sense of suffocation. Suffering is severe.[8]

By 1925 in this country a University of Pennsylvania physician had documented black lung symptoms and connected them to the coal dust to which miners are exposed daily.[9] Yet it was almost three decades before the disease was officially recognized anywhere in the United States and measures taken to prevent it. The initial preventive steps were minimal at best, involving little more than requirements, generally unenforced, for mine ventilation and sprinkler systems. Many a company doctor knew black lung when he saw it, but it could cost him his job if he diagnosed a miner's illness as black lung, or if he dared to suggest that black lung was a coal mine–related disease.[10] The federal government ignored the increasing numbers of disabled and dying miners, even while the British declared in 1943 that black lung was a compensable occupational disease. A decade later, in 1952, Alabama finally followed the British example. It took more than another decade for Pennsylvania and Virginia also to compensate their stricken miners. When at last the federal government moved in 1969, mine operators were given three years to lower dust concentration levels in their mines to three milligrams per cubic meter of air, a standard the British already had found more than adequate to prevent miners' pneumoconiosis.[11]

Enforcement is the Catch-22. And miners point out that enforcement is more important than ever since the switch from conventional mining of the coal face, in which dust can be more easily controlled, to the longwall system. Meanwhile, the use of heavy machinery such as the continuous miner has produced far heavier concentrations of dust in working areas. Miners claim that ventilation methods often can aggravate the problem by blowing dust across the men rather than pulling it away from work areas.

One miner described longwall mining as being far dustier than the old method of one or two men blasting and shoveling coal in a small area of face. The big machines, working along a wide coal face, tend to blow the dust back over the miners. The ventilation system meanwhile may also pull it across the men. The coal dust can be so thick that it is hard to see, and a miner working on the longwall will literally have to feel his way around. When the crew goes off shift, everyone knows they are the longwall crew because they are so much blacker than the other miners. Another miner who works the longwall told us, "There's no doubt that we're breathing the same dust that our fathers and grandfathers breathed."

In longwall mining a large block of coal is cut out along the seam. A line of men then works the cutting machines across the exposed seam. This is in contrast to conventional shaft mining, where one or two miners work at a face in an area small enough to be curtained off, where water is sprayed to settle the dust with reasonable efficiency. Both longwall mining and the heavy machinery now employed have created the need for a new technology in ventilation and in the use of water in the mines. Miners have been advised to use respirators to avoid some of the dust, but many have found them uncomfortable and difficult to work in. Thus enforcement of safe dust levels is obviously the crux of the present black-lung problem.

In the past it took about ten years of working in the mines for simple black lung to develop. Some physicians vitally interested in the problem believe that if present dust regulations are enforced, they will probably enable miners to work twenty to twenty-five years before indications of simple black lung appear. The miner suffering simple black lung may retain sufficient lung function to avoid disability. At this stage there are few symptoms other than coughing and expectoration of sputum, the same sort of symptoms an individual would display after years of smoking cigarettes. However, once simple black lung develops, continued exposure to coal dust poses the risk of complex black lung, and once that stage is reached, with its massive fibrosis of the lungs, removal from exposure to coal dust is of little help. Although there may be a leveling off of symptoms for a period, the disease tends to be progressive. Frequently, in Stage Three, black lung simply runs its course without pause. The victim can no longer function, not even at minor tasks.[12]

The history of black lung, like the history of safety in the mines, is one of systematic disregard for the miners who went underground to make King Coal. It is not a pleasant story, nor one that West Virginia or the nation can be proud of. Nor is it a pleasant thought that, with the exception of a handful of physicians such as Doctors Rasmussen, Buff, and Kerr, and Dr. H. R. M. Landis, the medical community turned a blind eye to such a pervasive occupational health problem. It can be said black lung was not easy to diagnose, because it affects the cardiac and respiratory systems, where similar symptoms are caused by other diseases such as asthma and emphysema; however, little was done in our own country, even after the British study was published and steps were taken in England to alleviate coal dust in the mines. As late as 1977, Representative John H. Dent of Pennsylvania testified before a congressional hearing that coal-company doctors still were not allowed to certify that a miner's disease was related to coal dust or that a miner had died from pneumoconiosis. Instant unemployment for the doctor could be and was the result of such a diagnosis.[13]

Sadly, it is estimated that over the years black lung has killed and disabled an estimated 365,000 miners. Neither miners nor their doctors can be certain that the hazard does not still exist, at least to some extent.

11
Miners and Medicine Today

King Coal may have been dethroned, but he is not dead. The old town crier's phrase, "The king is dead; long live the king," still holds true in Appalachia. Even so, both miners and medicine are not the same as they were a century or even forty or fifty years ago. For the most part, the coal camps are gone, and the company doctor and nurses with them. The strip mining that is devastating the mountains has replaced a good many of the deep mines now worked out or abandoned. Yet smoldering slag heaps and rusty streams in all-but-deserted hollows still testify to the industry's once-feverish activity, while aging men crippled in lung and bone are ample witness to the hard and dangerous task of bringing out the coal.

In his *Panorama of Fayette County*, John Cavalier related the poignant story of a seventy-year-old woman who lived on in a deserted coal camp, dwelling in a corner of an old and rotting company store on the banks of the New River, without electricity, plumbing, or neighbors. During the winter she would soak quilts and let them freeze, and then put them up as walls or as part of a roof to keep the wind out. She was dubbed "the crazy woman of New River Canyon" because she supposedly considered the river fish to be her relatives and believed that after dark they came up to the banks to pay her

113

a visit. But some people thought that she acted strange just to keep the abandoned coal camp to herself.

There was no road to the old camp, only the Chesapeake and Ohio railroad. She would walk the track to Thurmond, a mile or so away, to collect her pension as a miner's widow. After she became too old to make the trek, the C & O trainmen would throw off enough coal by the store for her stove as they went by, and every few weeks they left groceries out in front. Finally, one day they noticed that the food remained untouched. When they investigated, they found the old lady dead under the fallen beams of the ancient company store.[1] Like the silent donkey engines, the motionless conveyor belts, the rusting tracks, and rotting tipples, she was a final vestige of a once-thriving coal region. Many of the coal camps have been bulldozed from the landscape. Many more simply decay piece by piece, building by building. Some companies sold their houses to the miners, but few now resemble the original homes. Paint, new additions, and spruced-up yards and gardens have transformed the once-dingy buildings.

In those mines still operating, there have been some dramatic changes, not only in the new technology of the machinery, which makes the pick-and-shovel days seem akin to the age of the dinosaurs, but also in the miners themselves. Children no longer go down into the mines as trappers, or in the tipple as breaker boys, but women go down as miners. Once considered bad luck in the mines, they now don hard hats and carry lunch pails as their fathers and husbands did. Meanwhile, other things are little changed, such as the hazards miners face, even women. Barbara Angle, a miner in Keyser, West Virginia, testified to that: "Myself, I had the experience of being one of the first pregnant women to work underground. . . . I had no complications, but I know of various women who attempted to work the length of their pregnancies and had deformed children. . . . Theory is that coal dust [is] crossing the placenta, but it will take years to prove."[2]

After her child was born, Ms. Angle returned to the mines, working from 1975 until 1978 in all, when as she stated, "My career was cut short when a shuttle car crushed my right arm in 1978 I supposedly have 78 percent disability. But that's the Catch-22. Not fit enough to work, not hurt enough to get benefits."[3] Apparently, the rewards of hazardous work are still much the same.

114

Barbara Angle may not have been the first woman to go down into the mines. Resta Cheuvront tells of a trip she made in defiance of the "bad-luck" superstition:

> ... One night my husband (a miner for forty-three years, she says) was called to go inside and cut six places so men could work the next day. He tried very hard to get someone to go with him and could not get a man. I told him I would go and I did. I shoveled bug dust while he ran the machine and cut the coal. He turned the time in for his buddy so no one would even know his wife helped him so others could work the next day. That was our own big laugh for many years.[4]

The somewhat feudal system of the coal camp has also disappeared. Michael Large of Saint Charles, Virginia, miner for nine years, said his company has no company store, that he is covered by an insurance plan, and that if a miner is seriously injured, the federal government sends in inspectors. There is a medical center within five miles of the mine, and every miner working there is required to take rescue training (usually on his first day and lasting all day long), which includes cardiopulmonary resuscitation and other rescue maneuvers of all kinds.

Miners are also required to wear hard hats in designated areas, and where there is the possibility of methane gas buildup, they are required to don self-contained breathing apparatus. At Michael Large's mine there is in each section one person specially trained much like a paramedic, who usually has taken emergency medical training at the nearby community college. This person can be at the scene of an accident in the section within fifteen minutes, and if the accident is major, other paramedical people also will help out.[5]

Large went on to say that his grandfather was a miner and handloaded coal back in the 1930s, and that he now is suffering from black lung. In essence, concerns for the health and safety of miners altered only very slowly and, in all probability, only because of strong union prodding. When in 1983 amendments were proposed to change the Federal Mine Safety and Health Act of 1977, Joseph A. Main, a spokesman for the United Mine Workers of America, testified before a congressional committee that eliminating mandatory inspections of surface mines, cutting back on the training of miners, and reducing or eliminating the penalties for mine operators who violate the law are no answers to the continuing health and safety problems in the mines. The answer, he said, lay in stronger enforcement of existing laws.[6]

At the same 1983 hearing Senator Jennings Randolph of West Virginia made an equally strong statement against any amendment that would weaken the Mine Safety and Health Act. He said that mining was the most dangerous of all occupations in the nation and that, while mine explosions and fires are the most dramatic hazards, black lung is just as disastrous for miners. He too advocated stronger enforcement of mine safety and health laws.[7]

It should be noted that Senator Randolph was an original sponsor of the 1969 Federal Coal Mine Health and Safety Act, the first really adequate approach to regulating dust and gas in the mines as well as providing other safety measures. That law recognized the reality of black lung and the terrible toll of lives it has taken through the years—an estimated 365,000 coal miners disabled or dead. While the 1969 act's coverage was not complete, it provided the foundation for the more encompassing Federal Mine Safety and Health Amendment Act of 1977.[8]

While living conditions in the coalfields have improved, there are still problems and problem areas. In this era of modern sanitation, not all miners' homes have indoor plumbing, not all mountain creeks run free of sewage, not all houses are large enough for growing families, and there are still pockets of dismal poverty in Appalachia. Most of the coal companies either sold the old camp houses or tore them down, but those that were simply abandoned present sanitation problems for their few inhabitants because they are far back in remote hollows and a good distance from modern conveniences. Such areas also present problems for medical outreach programs, and as in other regions in the nation, family doctors are few and far between. That old custom of house calls is a thing of the past. Regional medical clinics are the wave of the future in the coalfields, but, as Shirley C. Rogers of Appalachia, Virginia, said:

They used to have a clinic or doctor's office in each mining camp. The doctors made house calls and nine times out of ten could get to a mining accident before the injured could be pulled out of the mine.

After unionization and the issuance of health cards by the UMWA, these offices closed; no more house calls. The injured had to be taken to a medical facility miles away.

About two years ago, they put nurse stations at a couple of the mine sites with RNs around the clock. They could treat minor injuries at these sites. They were financially better off but physically no better off.[9]

Many observers concur. The lack of adequate medical care in the coal country stems not only from the shortage of doctors—a condition probably present in most rural areas across the nation—but also from a shortage of up-to-date (and expensive) special medical equipment. While those company health cards provide miners and their families with health-care insurance, serious injury or disease means a trip to an urban area for specialized treatment or hospitalization. In addition, the medical clinics in coal communities are tied financially to the varying prospects of the coal industry. When the market for coal is poor, the clinics must curtail or even halt their services. Generally, however, since World War II the health care for miners and their families is very much better than that provided, if it was provided at all, in the coalfields before the war.[10] Some of the young may not think so, but those who lived through the era described in this book know so. And efforts are continuing to make health care even better, as Mrs. Hersel M. Perdue of Wayne, West Virginia, described:

Educational Seed for Physicians (ESP) is a medical student loan program designed to help promote and encourage family physicians to practice in West Virginia. It originated in 1978–1980 when the nearly 15,000 members to the West Virginia Federation of Women's Clubs raised over $60,000 for ESP. . . .

A student receiving an ESP loan promises to specialize in the field of Family Practice. Upon completion of his or her intern or residency training, he or she promises to practice in the State of West Virginia for five years.

Ms. Perdue goes on to say that if the recipient of the loan fulfills the promise, the loan need not be repaid. It is an exceedingly worthwhile program and should help fill the gap in rural health care.[11]

As the coal camp faded into the past, so did the coal-camp doctor. As it has elsewhere, medicine has taken some giant strides in the coalfields, with modern clinics and up-to-date physicians' offices. State-of-the-art hospitals have mushroomed in towns large and small. Yet, the medical community cannot afford to be complacent. There is still the kind of poverty among miners, former miners, and their families that breeds poor health and suffering. There are still mine accidents that cripple or kill. There is still black lung. There is still much to be done.

I know that my father's life as a coal-camp doctor must have been incredibly tough, but I am also certain that it had its rewards, for he brought medical care to people who otherwise would have had little or none. I am certain, too, that he left a legacy of gratitude back there in the hills and hollows, gratitude that many who have written me about their lives in the coalfields have voiced in their letters; and I am sure that he would have valued that above all.

Finally, let it be said that what was true for the coal camps and coal-camp doctors in West Virginia was also true in the main for the mining camps and camp physicians in other areas—Kentucky, Pennsylvania, Tennessee, the West—wherever coal was king.

Notes

1. King Coal

1. Bowen, *Pictorial Sketchbook of Pennsylvania*, 214–15.
2. Ibid., 214–15.
3. Ibid., 173.
4. Cavalier, *Panorama of Fayette County*, 119.
5. Bowen, *Pictorial Sketchbook of Pennsylvania*, 207.
6. Ibid.
7. Cheuvront to author, October 12, 1987.
8. Cavalier, *Panorama of Fayette County*, 119–20.
9. *Ibid.*, 405.
10. Caudill, *Night Comes to the Cumberlands*, 74.
11. Ibid., 75.

2. The Coal Camps

1. Caudill, *Night Comes to the Cumberlands*, 113.
2. McGill, *Welfare of Children*, U.S. Children's Bureau Publication 117, 11.
3. Ibid., 11.
4. President's Commission on Coal, *The American Coal Miner* (1980), 32.
5. McGill, *Welfare of Children*, U.S. Children's Bureau Publication 117, 14–16.
6. Quoted in President's Commission on Coal, *The American Coal Miner*, (1980), 32–33.

7. Littleton, "Memories of a Coal Camp Physician," 14–15.
8. Ibid.
9. Bryant to author, September 3, 1987.
10. Ibid.
11. Akers to author, August 17, 1983.
12. Taylor to author, August 22, 1983.
13. Blount to author, August 8, 1983.
14. Lilly to author, July 28, 1984.
15. Cheuvront to author, September 24, 1987.
16. McGill, *Welfare of Children*, U.S. Children's Bureau Publication 117, 37–46.
17. Halstead to author, August 1, 1983.
18. Jim Blankenship to author.
19. Lilly to author, July 28, 1984.

3. The Miner's Lot

1. Harrington, *Welfare Problems of the Mining Industry*, Report of the U.S. Bureau of Mines, May 1939, 5.
2. Ibid., 4.
3. Martin to author, September 7, 1983.
4. Prather to author.
5. Craddock to author, August 31, 1983.
6. Tams, *Smokeless Coal Fields of West Virginia*, 36.
7. Ross, unpublished "Memoirs."
8. Ibid.
9. Ibid.
10. Henderson to author, October 17, 1983.
11. Radford to author, August 2, 1983.
12. McGill, *Welfare of Children*, U.S. Children's Bureau Publication 117, 38.
13. Ibid., 39.
14. Rosner and Markowitz, *Dying for Work*, 71–72.
15. President's Commission, *The American Miner*, 78.
16. Palmer to author, December 31, 1983.
17. Ibid.

4. Death and Danger in the Mountains

1. Tams, *Smokeless Coal Fields of West Virginia*, 50.
2. Ibid., 50.
3. Smith, *New York American*, December 7, 1907, republished in *West Virginia Hillbilly*, December 31, 1987.
4. Smith, *New York American*, December 8, 1907, republished in *West Virginia Hillbilly*, January 14, 1988.
5. Wells to author, August 5, 1983.
6. Rice, *West Virginia*, 237.

7. May to author, August 11, 1983.
8. Green interview, November 17, 1984, Powellton.
9. Arneach to author.
10. Dillon, *They Died in Darkness*, 8.
11. Ibid., 10.
12. Ibid., 1.
13. Cavalier, *Panorama of Fayette County*, 371.
14. Donnelly, *Montgomery Herald*, April 8, 1982.
15. Cavalier, *Panorama of Fayette County*, 403.
16. Ibid., 231.
17. Ibid., 222.
18. Ibid., 127.

5. Old-time Medicine

1. Duffy, *The Healers*, 17–18.
2. Ibid., 26–30.
3. Dolan, *Nursing in Society*, 230.
4. Ibid., 230.
5. Duffy, *The Healers*, 151–54.
6. Ibid., 22–25, 31–35.
7. Ibid., 58.
8. Summers to author.
9. Taylor to author, August 22, 1983.
10. Green and Bennett, "Catfish," *Goldenseal* 7, no. 3 (Fall 1981): 49–51.
11. Fisher to author, September 6, 1987.
12. Jones to author, December 1, 1984.
13. Gray, "Remembering the Past," *West Virginia Hillbilly*, July 21, 1985.
14. Smith, *Patent Medicine*, 2–19.
15. Duffy, *The Healers*, 294.
16. Fisher to author, September 6, 1987.
17. Tompkins to author.
18. Lilly to author, July 28, 1984.

6. The Coal-Camp Doctor: I

1. Werner, *Big Doc and Little Doc*, 13–14.
2. McGill, *Welfare of Children*, U.S. Children's Bureau Publication 117, 48–49.
3. Mayes to author, February 5, 1988.
4. Werner, *Big Doc and Little Doc*, 13–14.
5. Jones to author, September 11, 1987.
6. Clark to author.
7. Littleton, "Memories of a Coal Camp Physician," 7.
8. Chandler to author.

9. Fisher to author, September 6, 1987.
10. Weatherford to author, August 8, 1984.
11. Werner, *Big Doc and Little Doc*, 78.
12. Lilly to author, July 28, 1984.
13. Lambert to author.
14. Werner, *Big Doc and Little Doc*, 120.
15. Blount to author, August 8, 1983.
16. McGill, *Welfare of Children*, U.S. Children's Bureau Publication 117, 49.
17. Harper to author.
18. Jones to author, August 2, 1983.
19. Ibid., December 1, 1984.
20. Ibid., December 1, 1984.

7. Coal-Camp Doctor: II

1. Taylor to author, August 22, 1983.
2. McGill, *Welfare of Children*, U.S. Children's Bureau Publication 117, 47.
3. Soulsby to author.
4. Ibid.
5. Williams to author, January 27, 1985.
6. Linsky to author, August 1, 1983.
7. President's Commission on Coal, *The American Coal Miner*, 114.
8. Marshall Werner, telephone interview, Asheville, N.C.
9. Harper to author.
10. Soulsby to author.
11. Taylor to author, August 22, 1983.
12. Linsky to author, August 1, 1983.
13. Littleton, "Memories."
14. Ibid.
15. "I recall one house call": Ibid.
16. "One thing I would like to say": Benger to author, December 17, 1984.
17. "I can remember how for so long": Lilly to author, July 28, 1984.
18. Hopkins to author, October 27, 1987.

8. Miners' Hospitals

1. Penn, "Henry D. Hatfield and Reform Politics," Ph.D. diss., Emory University, 1973, 1–3.
2. Ibid., 4.
3. West Virginia State Board of Control, *Miners' Hospital Number One: First Biennial Report* (1910), 108, 113–16.
4. *Welch* (W.Va.) *Daily News*, September 21, 1926.

122

5. West Virginia State Board of Control, *Miners' Hospital Number One: First Biennial Report.*
6. Ibid.
7. Henry Hatfield to John L. Lewis, February 4, 1955, Hatfield Papers, West Virginia University, Morgantown.
8. West Virginia State Board of Control, *Miners' Hospital Number One: First Biennial Report,* 125.
9. Ibid.
10. Cox, *Goldenseal* 7, no. 3 (Fall 1981): 39.
11. Gray, "Remembering the Past," *West Virginia Hillbilly,* September 15, 1984.

9. The Nurses

1. Dolan, *Nursing in Society,* 162–63, 137.
2. Ibid., 155–59.
3. Ibid., 173–76.
4. Ibid., 191–94.
5. West Virginia State Board of Control, *Miners' Hospital Number One: First Biennial Report,* 131.
6. Fisher to author, September 6, 1987.
7. Ibid.
8. Ibid.
9. Ibid.
10. Frazier, *Healing and Religious Faith,* foreword.
11. Ibid.
12. Nyden, "Mabel Gwinn, New River Nurse," *Goldenseal* 7, no. 3 (Fall 1981): 34–35.
13. McKean to author, February 6, 1988.
14. Ibid.
15. Ibid.
16. Fisher to author, October 27, 1987.

10. Coal Miner's Nightmare: Black Lung

1. Lynch, "Tragedy at Hawk's Nest Tunnel," *West Virginia Hillbilly,* November 6, 1982.
2. Ibid.
3. Kerr, "Black Lung," *Journal of Public Health Policy,* March 1980. After twenty years of working for the United Mine Workers Association Welfare and Retirement Fund, Dr. Kerr was an expert in all aspects of coal workers' pneumoconiosis and a longtime advocate of federal recognition that black lung is an occupational disease that requires not only compensation but also treatment.
4. Ibid., 55.
5. Ibid., 55.
6. Ibid., 56.

7. Spittell, *Clinical Medicine*, 19–21.

8. U.S. Department of Health, Education, and Welfare, Publication 74-488, pp.10–13.

9. President's Commission on Coal, *The American Coal Miner*, 120–22.

10. Kerr, "Black Lung," *Journal of Public Health Policy*, 8.

11. Ibid., 11.

12. Spittell, *Clinical Medicine*, 21.

13. President's Commission on Coal, *The American Coal Miner*, 122.

11. Miners and Medicine Today

1. Cavalier, *Panorama of Fayette County*, 111.

2. Angle to author,

3. Ibid.

4. Cheuvront to author, October 12, 1987.

5. Large to author.

6. U.S. Congress, Subcommittee on Labor, *Federal Mine Safety and Health Amendments of 1983*, hearing before Subcommittee on Labor, July 26, 1983, 66.

7. Ibid., July 16, 1983, Randolph, 96–97.

8. Ibid., 100.

9. Rogers interview.

10. President's Commission on Coal, *The American Coal Miner*, 781.

11. Perdue to author, September 3, 1987.

Bibliography

Books

Bowen, Eli. *Pictorial Sketchbook of Pennsylvania.* Philadelphia: William Bromwell, 1853.

Caudill, Harry M. *Night Comes to the Cumberlands.* Boston: Little, Brown and Company, 1962.

Cavalier, John. *Panorama of Fayette County.* Parsons, W.Va.: McClain Printing Company, 1985.

Cox, William E. *Life on the New River: A Pictorial History of the New River Gorge.* Eastern National Park and Monument Association, 1984.

Dillon, Josephine A., M. Louise Fitzpatrick, and Eleanor Krohn Hermann. *Nursing in Society: A Historical Perspective.* Philadelphia: W. B. Saunders Company, 1983.

Duffy, John. *The Healers: A History of American Medicine.* 1976. Champaign: Illinois Press, 1979.

Frazier, Claude A. *Healing and Religious Faith.* Philadelphia: United Church Press, 1974.

Homer, Morris L. *The Plight of the Bituminous Coal Miner.* Philadelphia: University of Pennsylvania Press, 1934.

Rice, Otis K. *West Virginia: A History.* Lexington: University Press of Kentucky, 1985.

Rosner, David, and Gerald Markowitz. *Dying for Work: Workers' Safety and Health in Twentieth Century America.* Bloomington: Indiana University Press, 1987.

Miners and Medicine

Shryock, Richard H. *Medicine and Society in America, 1660–1860.* New York: New York University Press, 1960.

Smith, Elmer L. *Patent Medicine: The Golden Days of Quackery.* Lebanon, Pa.: Applied Arts Publishers, 1973.

Spittell, John A. *Clinical Medicine.* Philadelphia: Harper and Row, 1982.

Tams, W., Jr. *The Smokeless Coal Fields of West Virginia.* Morgantown: West Virginia University Library, 1963.

Werner, Harry R. *Big Doc and Little Doc.* As related to Emily Dane Werner. Parsons, W.Va.: McClain Printing Company, 1980.

Articles

Cox, William E. "McKendree No. 2." *Goldenseal* (West Virginia Division of Culture and History) 7, no. 3 (Fall 1981).

Donnelly, Shirley "Slate Falls Big Killer in Fayette Mines." *Montgomery* (W.Va.) *Herald,* April 8, 1982.

Gray, Clyde "Remembering the Past 100 Years and Then Some." *West Virginia Hillbilly* (Richmond, W.Va.), September 15, July 21, July 24, 1984.

Green, Ted, and Allen Bennett. "Catfish: Portrait of an Herb Doctor." *Goldenseal* 7, no. 3 (Fall 1981).

Kerr, Lorin E. "Black Lung." *Journal of Public Health Policy* (Department of Occupational Health, United Mine Workers of America), March 1980.

Morgan, W. K. C. "Occupational Pulmonary Disease." In *Clinical Medicine,* edited by John A. Spittell, Jr. Philadelphia: Harper and Row, 1982.

Nyden, Paul J. "Mabel Gwinn, New River Nurse." *Goldenseal* 7, no. 3 (Fall 1981).

Smith, Langdon: "Disaster at Monongah." *New York American,* December 7, 1907; reprinted in *West Virginia Hillbilly,* December 31, 1987.

———. "Monongah." *New York American,* December 8, 1907; reprinted in *West Virginia Hillbilly,* January 14, 1988.

Government Publications

Harrington, D. *Some of the Welfare Problems of the Mining Industry and What the Bureau of Mines has Done About Them.* Report of the U.S. Bureau of Mines, Department of Interior, May 1939.

McGill, Nettie P. *The Welfare of Children in Bituminous Coal Mining Communities in West Virginia.* U.S. Children's Bureau, Department of Labor, Publication 117. 1923.

President's Commission on Coal. *The American Coal Miner.* Washington: Government Printing Office, 1980.

U.S. Congress. Subcommittee on Labor. *Federal Mine Safety and*

Bibliography

Health Amendments of 1983: Hearings, July 26, 1983. Washington: Government Printing Office.

West Virginia State Board of Control. *Miners' Hospital Number One: First Biennial Report,* 1910.

Unpublished Materials

Thesis

Penn, Neil Shaw. "Henry D. Hatfield and Reform Politics." Ph.D. diss., Emory University, 1973.

Memoirs

Littleton, L. R., M.D. "Memories of a Coal Camp Physician." In the author's possession.

Ross, Howard, Sr. "Memoirs." Courtesy of Margaret Spriggs.

Letters to Author

Akers, Estella, Hinton, W.Va., August 17, 1983.

Angle, Barbara, Keyser, W.Va.

Arneach, Richard E., Sylva, N.C.

Benger, Dorlene, Ansted, W.Va., December 17, 1984.

Blankenship, Jim, Davin, W.Va.

Blount, H. Neil, Salinas, Calif., August 8, 1983.

Bryant, Katherine, Liverpool, W.Va., September 3, 1987.

Chandler, Russell, Paintsville, Ky., July 29, 1984.

Cheuvront, Resta, Murrells Inlet, S.C., September 24, October 12, 1987.

Clark, Claude, Williamson, W.Va., February 12, 1986.

Craddock, Madeline, St. Albans, W.Va., August 31, 1983.

Fisher, Pauline, Union, W.Va., September 6, September 24, September 24, October 7, October 27, 1987.

Halstead, Mrs. E. Houston, Scott Depot, W.Va., August 1, 1983.

Harper, Ruth A., Procious, W.Va.

Henderson, Earl D., Robinette, W.Va., October 17, 1983.

Hopkins, Evelyn A., Bluefield, W.Va., October 27, 1987.

Jones, June, Charleston, W.Va., August 2, 1983, December 1, 1984.

Jones, Lillian, Huntington, W.Va., September 11, 1987.

Lambert, Elizabeth, Rainelle, W.Va., May 2, 1988.

Large, Michael, St. Charles, W.Va.

Lilly, Lillian, Huntington, W.Va., July 28, 1984.

Linsky, Rose, Charleston, W.Va., August 1, 1983.

Littleton, L. R., Jr., M.D., Webster Springs, W.Va., September 4, 1987.

McKean, Eva Ruth, Marlinton, W.Va., February 6, 1988.

May, Alice, Clifftop, W.Va., August 11, 1983.

Martin, Hazel, Oak Hill, W.Va., September 7, 1983.

Mayes, Charles M., Deerfield, Fla., February 5, 1988.

127

Miners and Medicine

Palmer, O. E., Cedar Grove, W.Va., December 31, 1983.
Perdue, Mrs. Hersel M., Wayne, W.Va., September 3, 1987.
Prather, Bonnie, Charleston, W.Va.
Radford, Lucille, Kimberly, W.Va., August 2, 1983.
Soulsby, Paul, M.D.
Summers, Jack C., Charleston, W.Va.
Taylor, Martha E., Meadow Bridge, W.Va., August 22, 1983.
Tompkins, Margaret E., Lansing, W.Va., September 21, 1987.
Weatherford, Elizabeth M., True, W.Va., August 8, 1984.
Wells, Ahlena L., Huntington, W.Va., August 5, 1983.
Williams, Robert L., South Charleston, W.Va., January 27, 1985.

Interviews

Green, Samuel, Powellton, W.Va., November 17, 1984.
Rogers, Shirley, Appalachia, Va., Asheville, N.C., August, 1987.
Werner, Marshall, M.D., telephone interview, Asheville, N.C.

Index

Index

Index